Day Range Poultry

Every Chicken Owner's Guide
to Grazing Gardens and Improving Pastures

Including the Management of Breeder Flocks,
Egg Handling, Incubating Secrets, Hatchery Effi-
ciency, Building Shelters, Marketing,
Advertising, Soils Regeneration,
Processing Poultry Humanely and Efficiently,
and Much, Much, More!

Library of Congress Cataloging-in-Publication Data

Lee, Andrew W., 1948-

Day Range Poultry : every chicken owner's guide to grazing gardens and improving pastures / by Andy W. Lee. and Patricia L. Foreman.

– 1st ed

p. 320 cm.

Includes bibliographical references and index.

Preassigned LCCN: 92-076194

Preassigned   ISBN 0-9624648-6-4

1. Poultry. 2. Organic farming. 3. Sustainable Agriculture. 4. Waste products as fertilizer. 5. Community food self-sufficiency

I. Foreman, Patricia L. II. Title

SF487.L44 2001   636.5

QB100-900   *22.00   Ingram 1/06*

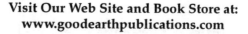

Published in the United States by
Good Earth Publications, Inc.

Visit Our Web Site and Book Store at:
www.goodearthpublications.com

Printed in the USA

Acknowledgments

Thanks to **Paula and Steve Weiss-Martin, Matt Jones, Joel Bauman, Jeff Ishee, Ben Rippy, Amanda Gross, Paul Garp, Anna Haupt, Annabel Por,** and all the others who have worked at Living Earth Organic Farm for their parts in helping with the research and development that is the backbone of this book. Also a big thanks to **Healthy Foods Market, Southern Inn, Main Street Market, and all our customers** in Rockbridge County, Virginia for their patronage, support, and appreciation of wholesome and humanely raised food.

And Special Thanks To:

Manal Stulgaitis, for being just who you are, bringing your special light to darkened areas of the world.

Michelle duBois Peterson, for using her Virgo midpoints to mold the manuscript into a grammatically correct work-of-art.

Tim Shell, for many long thought-provoking discussions, and for helping to pioneer the pasture peeps. Keep thinking, Tim!

Joel, Theresa, Daniel and Rachel Salatin, for their friendship, laughter, support, and pioneering work in pasture poultry and family-friendly farming.

David and Maggie Cole, for operating Sunnyside Farms, and for their contributions to conservation and organic farming.

Donald Bixby, Majorie Bender, and the staff of the American Livestock Breeds Conservancy, for the service they are providing in preserving our heirloom livestock and poultry.

Dave & Jeff Mattocks, for making organic feed supplements and soil amendents available through Fertrel.

Phil Laughlin, cover design, and creative computer wizardry.

Michael Plane, our Australian friend who introduced us to British Electronet fencing.

The Coveted

Pullets Surprise

for

Poultry Pioneers

Hall of Fame

The Staff of
The American Livestock Breeds Conservancy
for their work in preserving heritage breeds of poultry."

Herman and Linda Beck-Chenoweth
for publishing the Range Poultry Forum.

Gail Damerow
for her many books and articles on keeping small poultry flocks.

Glen Drowns
for his contributions to poultry conservation at the Sand Hill
Preservation Center.

Anne Fanatico
for her research and contributions to
poultry on pasture.

Bruce Johnson
for creativity in chicken tractor design.

Andy Lee and Patricia Foreman
for helping to integrate millions of chickens into
gardens and pastures across the globe.

Steve Muntz
for his work with poultry cooperatives and
mobile poultry processing units.

Joel Salatin
for his paradigm-changing books, inspirational
lectures, and supporting family farming.

Tim Shell
for promoting the pasture peeps and other contributions to the
evolving day range poultry systems.

Kathie Thear
for her books and contributions to free range poultry in England.

David White
pioneer grower and breeder of pasture poultry.

And Every Poultry Owner
who cares for, and respects their birds
and the service they give to humanity.

About the Authors

Andy Lee was born and raised in Missouri, He is a well respected and an internationally known speaker, writer, teacher, and practitioner of what he has come to call "holistic enterprise farming and gardening". His work encompasses sustainable agriculture, practical permaculture, conservation housing, rural economic development, and cottage-industry strategies. His articles have appeared in dozens of publications including *Organic Gardening, Country Journal, Newsweek, Vegetarian Times, Small Farm Today, Acres USA, BackHome Magazine, Permaculture Drylands Journal,* and other sustainable living magazines.

He is the co-author of *Backyard Market Gardening: The Entrepreneur's Guide to Selling What You Grow.* This book has been the flagship book for helping to establish farmers' markets across the US. Andy and Pat also wrote the best-selling book *Chicken Tractor: the Permaculture Guide to Happy Hens and Healthy Soil,* and *A Tiny Home to Call Your Own: Living Well in Just Right Houses.*

Andy has over 30 years experience as a forester, market gardener, ecological home builder and designer, and business owner. He helped rewrite the Massachusetts Forest Cutting Practices Act as a model to promote sustainable forestry in New England. He has a Diploma of Permaculture Design from the Permaculture Institute in Australia. He has received numerous awards, including the Excellence in Agriculture Award from the Renew America Foundation, the US Mayors' Award for the Fight against Hunger, the Silver Spade Award from the Janus Institute, and the coveted Pullets Surprise for Poultry Pioneers.

Patricia Foreman was born and raised in Indiana. She studied animal science (genetics and nutrition) and Pharmacy at Purdue University and graduated with degrees from both colleges. Continuing her studies at Indiana University, Pat earned a Master's degree in Public Administration with majors in Health Systems Administration and

International Affairs. While attending Indiana University, she was selected for a prestigious Fulbright Fellowship.

After graduating in 1976, she went to Washington DC as a Presidential Executive Management Intern under the Carter Administration. She then worked as a Science Officer for the United Nations in Vienna, Austria. Later, Pat joined Management Sciences for Health, a nonprofit firm. For over a decade, she worked as a consultant on primary health care systems and family planning programs in over 30 developing countries. She has been the course director for many international and trans-cultural courses and has extensive experience in adult learning and training. Her interests keep bringing her back to the role nutrition plays in health care — nutrition that begins with healthy soils and animals, that produce healthy foods.

In 1991, Pat and Andy took their first permaculture design training course with Bill Mollison, the founder of the worldwide permaculture movement. This was followed by village design and teacher training courses with Australians Max Lindegger and Lea Harrison in 1992. Pat and Andy have worked together on numerous permaculture designs including Kripalu Institute in Massachusetts, and Findhorn Community in Scotland. For nearly five years they helped develop the Intervale Community (CSA) Farm and the non-profit Intervale Foundation in Burlington, Vermont.

In 1989 they created the "not-for-profit-only" Good Earth Publications to produce books that promote sustainable life styles, permaculture, eco-agriculture, and community self-reliance.

Pat and Andy have retired from farming. They are now turning their attention to the problems of valuable farm land being used to grow the ultimate crop: houses. Their newest book, *A Tiny Home to Call Your Own; Living Well in Just Right Houses* is about conservation development, cluster housing, tiny houses, and ecological and healthy home construction. They are owners of the Tiny House Company, Inc., and owners and developers of GreenWay Neighborhood near Lexington, Virginia. Greenway Neighborhood is designed using permaculture and ecological design precepts.

The authors

Patricia Foreman and Andy Lee

Donate

All of their Royalties

after taxes from the sale of this book

to

The Gossamer Foundation

for educational, charitable, and
environmental restoration purposes.

For information about
grants, scholarships, internships, and awards

visit

GossamerFoundation.com

Dedicated to

Brian William Lee

and

Christopher Andrew Lee

Two fine young men who live on in spirit.

They serve as our inspiration,

dedication, hope, and love

Some people, sweet, attractive,
and strong and healthy,
happen to die young.
They are masters in disguise,
teaching us about
impermanence.

Dali Lama
The Path to Tranquility

Table of Contents

Recipe for Day Range Poultry

The name "Day Range" comes from allowing poultry to range freely within a grassy paddock during the daytime and securing them inside a weather-proof, predator-proof shelter at night. It is easy to do.

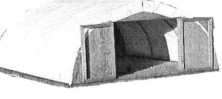

1. Make a portable chicken tractor with a floor that's on skids. The floor is a key component because it keeps the birds off wet ground and lets you move the pen without crushing or injuring any birds. We prefer the hoop-house model. Read why later.

2. Add deep bedding, which you can collect, compost, and use later in your garden to grow the best vegetables and fruits ever.

3. Add field waterers and feeders. See what we use and how to make your own cheaply in Chapter 7.

4. Add poultry: broilers, laying hens, or turkeys.

5. Surround the entire building and some pasture with electric netting to provide protection. Learn the shocking truth about how to use electric fence.

6. In the evening, chase the birds in and close the doors to protect the birds from night time predators and bad weather. During the morning chores, open the doors and let the birds out.

Here are the results: Your birds have access to pasture, sunlight, fresh air, and can freely exercise. There is almost zero mortality loss due to weather or overcrowding. And, there are no injured birds from moving the pens. When new pasture is needed, you simply move the electric netting to new graze, or move the building and reset the electric netting.

The Difference Between
Day Range Poultry and Chicken Tractor

The major differences in these two books are summarized below. Each book contains unique material and new information. The entire tables of contents of both books can be seen on the Good Earth Book Store website at:
www.GoodEarthPublications.com

Day Range Poultry	Chicken Tractor
Directed at larger, commercial pasture poultry growers.	Directed at small-scale homestead poultry growers.
Farm, pasture, market garden, and multi-species grazing.	Kitchen garden and homestead permaculture design.
Commercial processing of hundreds or thousands of birds, including USDA facilities.	Table top processing for the home and a few customers.
Details about many kinds of larger shelters (with bottoms) that are moved by motorized vehicles (tractors, ATV, etc.) or by hand.	Smaller, bottomless shelters with pop-holes for good weather use and for building raised garden beds. Moveable by hand.
Covers incubation, hatchery management, and battery brooding.	Includes straw-bale chicken shelters for winter housing.
Detailed information and lessons-learned about electric poultry fencing.	Information on movable fencing, including bamboo panels.
Detailed information about raising turkeys and poultry breeder flocks.	Main focus on chickens (layers and broilers).
Bulk organic fee ration formulas by the ton for the commercial grower.	Details about feed supplements.
Stresses why a poultry growers' cooperative would benefit all growers.	Centers around farmer's markets and direct-sale customers.

Foreword by Gene Logsdon

Having made what the smart money said was a foolish spectacle of myself a few years ago by saying on the radio that the age of the backyard chicken was cycling around again, I was delighted to learn that Pat Foreman and Andy Lee were publishing their *Day Range Poultry*. I was similarly delighted with their first book, *Chicken Tractor*, for the same reason, especially after it became a best seller among all of us forward-to-the-landers. And I absolutely giggled with elation when Martha Stewart decided to get on the backyard chicken bandwagon just as artist Marsha Tudor and poet-novelist Wendell Berry had been for years. Now if we could just get Oprah...

Day Range Poultry is an immensely practical book. It refines and expands to a commercial level the techniques presented in *Chicken Tractor*. Whereas *Chicken Tractor* is aimed at the gardener level with a few backyard hens, *Day Range* is aimed at the small scale commercial grower of chickens and turkeys. Both books contain original information and are a valuable additions to any library.

The term "Day Range" means the animals range or graze during the day within protective electric fencing, and are secured at night in shelters safe from predators and inclement weather. It was necessary to find a term other than "free range" anyway, because mega-growers have, in another example of the rising, modern Tower of Babble, usurped the phrase "free range" to mean little more than thousands of chickens being able to range "freely" inside a chicken factory. The "chicken tractors" become, in this new book, more substantial and utilitarian, although still movable. These and other refinements answer the basic problem that has always plagued the practice of allowing chickens to roam freely in fields: too many predators.

My first test of a new book on domestic farm animals is to look for information on fencing. If there is little on that subject, I know the authors are not putting proper emphasis on the first step to success with farm animal enterprises - keep your animals on your own property. *Day Range Poultry* does have a whole chapter on fencing and gives details about electric fencing techniques found in few books. Reading the authors' instruction leaves no doubt in my mind that they actually do what they are writing about. I refuse to read any book about farming written by someone not doing what they are experting about, especially if written by university professors.

My second test is to note whether the book has detailed, honest, and practical information on predator control. Wild animals have become the number one problem in small farm and garden enterprises, especially in poultry. With environmental emphasis on endangered species, society is not generally aware that many species of birds and mammals are not only no longer endangered, but are actually exploding in population, especially in and around suburban areas.

Deer, raccoons, rats, rabbits, squirrels, skunks, coyotes, groundhogs, and Canada geese, to name only the worst, are becoming very real problems. In an agrarian society, control of wildlife populations was managed as adroitly as control of domestic animal population but today, cut off from the realities of nature, many urbanites are unable, until wildlife threatens them, to grasp population dynamics.

Most how-to book publishers soft-pedal predator control or require their writers to avoid it altogether. Lee and Foreman address the problem directly. They discuss, in detail, how pests can be controlled without killing them most of the time but they are not afraid to say that sometimes killing predators is the necessary response.

Once in Real Goods News, I wrote that predator control, including trapping and killing, was part of my daily routine. If it weren't, I might as well quit trying to raise food and move to a high-rise and self-righteously condemn hunters while eating food that I wouldn't have if farmers and hunters did not control wildlife populations.

Oh, what consternation hit the Letters to the Editor column in the next issue. About twenty sidewalk environmentalists called me a craven, bloodthirsty, barbarian. But in the issue following that outburst, more than fifty gardeners, farmers, and experienced environmentalists supported me.

I am fond of the story of deer loving suburbanites near Lyme, Connecticut who successfully fought for the prohibition of deer hunting in their suburban areas until Lyme's Disease showed up there (hence the name) and was traced to deer. Then some of the same people who had opposed hunting wanted <u>all</u> the deer killed.

Foreman and Lee choose the middle path, and in so doing do the reader a great service. If you are going to raise poultry, you had better understand that rats, raccoons, foxes, coyotes, skunks, minks, weasels, cats, dogs, and hawks, to name a few, are going to get your chickens if you don't watch out. Life is a great dining table around which living beings all sit, eating and being eaten. Until everyone understands that, environmentalism is just an airy word with too many syllables.

I also like *Day Range Poultry* because the material is presented in a lively and humorous way. Let's face it: there must be half a hundred books on chickens available and most of the information contained in them is generic, common to all books about chickens. There is nothing more boring than trying to read that information presented in a sober, humorless way. Foreman and Lee delight us with humor, which often gives better insights on instruction than sober-sided accounts of putting tab A into slot B. For example, as the authors tell us how to build a Day Range shelter they also make the point, with a simple cartoon, that "Some shelters become box kites in high winds." There are dozens of such funny, right-to-the-point drawings and insightful photos that makes this book a delight to read.

But the main significance of *Day Range Poultry*, and the whole library of books and manuals of instruction that are now being published on small-scale food production, is that they defy and dismay the terribly dangerous mergers and acquisitions now underway in the trend toward monopoly in the food business. Large mega-companies make no secret of their intention to vertically integrate and gain control of the food business. From "seed to shelf" says Cargill; from "dirt to dining table" says Dow. Big business has decreed that you will eat genetically-engineered, chemicalized, and irradiated factory food or else.

But the big food monopolies don't understand us. As long as we have backyards and small farms, there is no way in heaven or hell that they can compete with us. They simply can't produce in their factories the eggs and fried chicken that I and my neighbors are eating as cheaply, or with as much quality as we can in our backyards. It is for that reason, contrary to what the savants of economics are so blindly predicting, that small scale food production enterprises are going to flourish now, and forever more. Books like *Day Range Poultry* are leading the way.

Gene Logsdon, Prolific Author
The Contrary Farmer.
Wildlife in the Garden: How to Live in Harmony with Deer, Raccoons, Rabbits, Crows, and other Pesky Creatures.
The Man Who Created Paradise: A Fable.
Gene Logsdon's Practical Skills: A Revival of Forgotten Crafts, Techniques, and Traditions.
Homesteading: How to Find New Independence on the Land.

Introduction, by Tim Shell

I remember standing in the field with Andy Lee looking at his Day Range setup and listening to him muse about the benefits of the Day Range model, and how it builds upon the concepts he and Pat wrote about in their book *Chicken Tractor: The Permaculture Guide to Happy Hens and Healthy Soil*.

I think of Andy's musing as evolution - improving on the good. Improvement comes marginally. The tragedy comes when one generation refuses to admit their mistakes and insists on continuing with an obsolete model, forcing the next generation into the same mistakes. It is only by doing the noble thing and acknowledging a better way that we help people improve on what is. I commend Andy for being able to do this.

There are a lot of people who simply have too much invested to be willing to change for the betterment of those who will follow. I remember watching Joel Salatin shift from the pasture pens to the Day Range portable netting model for Polyface's layer operation. He just stood there looking at all the 30-some layer pens he had built over the years. It's the same story – progress by evolution and experience. There is always a better way.

Both men were looking toward the future with a better model in their hand and a new vision of what could be done. There is nothing ignoble about saying, "I've found a better way," even if you developed the first way. Why keep propagating and promoting a thing you have discovered is inferior just to save face? I find something admirable in the people who can step up to the next level even though they invented the last one everybody is standing on.

It is one of the strange privileges of leadership to be able to pull others up to the next level by standing on the remains of your last discovery.

I know that some of the things that motivated Pat and Andy were their experiences with pastured broilers and feedback from other producers who were struggling to make it work. They saw a lot of realities were being ignored. It was their deep honesty and compassion they had for their birds that cleared their eyes.

The point I want to make is that Pat and Andy wanted to promote a model that retains the benefits of the pen model while eliminating the drawbacks. They achieved those goals very well. We criticize the confinement industry for being inhumane, but Pat and Andy were seeing and experiencing certain situations with pastured pens that were inhumane and intolerable as well.

When we think of pen pulling, I'm sure we all have visions of sunny days with warm June temperatures and balmy breezes, crickets chirping and broilers grazing contentedly on clover. That would be the photo on the front of our advertising brochure, right? We would never advertise with a photo of us out in the field at midnight in April in a cold rain, shivering, bleary eyed, and crawling around in the wet grass trying to move wet and bedraggled four-week old chicks up onto hay pads. Nor the photo of the wet chicks that didn't make it, or the carcass of the predator kill. The worse the weather, the harder you work to help the birds survive. But one is just as real as the other.

If you have fair weather all the time, then bottomless pen pulling might hold its own from an environmental perspective. The reality is that we do have to deal with bad weather. And the more you get into pastured poultry, the more you are driven to push the limits of the season in spring and fall in order to improve the profitability of your operation, which brings you face-to-face with more bad weather. This is where Day Range really shines.

Think about all the things we don't like about confinement poultry: over-crowding, de-beaking, fecal dust, ammonia fumes, artificial light, few bugs and worms, and no grass. But we have to admit, at least confinement poultry has climate control, and at least the birds don't shiver in the cold, stand in the rain, bake in the heat, or get eaten by owls and foxes.

Day Range is designed to provide all the benefits of the pens and some of the plus side of confinement. If you get three weeks of rainy weather in April and move the cage from wet grass onto more wet grass then how good can that be for your flock? How does it affect grow out? How does it affect your bottom line? With Day Range you can open the door and let the birds decide if they want to venture out or not. Your birds will like Day Range because you have already provided for the worst case scenario weather-wise.

In short, Day Range gives the birds more freedom to express their "chicken-ness" and they make full use of it. This is not to say that Day Rangers don't like to lounge around like nest-potatoes – they do. But they are also more active and have better muscle tone from the added exercise and freedom-of-movement they are allowed to have in the pens. This gives us a cleaner, healthier chicken.

I applaud Pat and Andy for their thoughtful, paradigm-changing work. This book is a testimony to their hard work and hands-on contribution to the farmers of today and tomorrow.

God bless you both, and thank you!

Timothy Shell
Master Farmer and Pasture Peeps Breeder
Burnsville, Virginia, USA

What is Day Range?

My dad was a quiet, pragmatic, unemotional man, not given to overstatement or embellishment. So it came as no surprise to me one day, when I had waxed long and eloquently about the color and graceful lines of my new barn roof, that he quietly gazed of in the direction of the barn and simply said, "Yeah, but will it hold water?" And when you get right down to it, that's all we can really expect from a good barn roof, and it is an excellent question for any new idea – to test if it will simply "hold water".

That is a particularly appropriate story as I stand here and introduce you to yet another model for raising poultry in gardens and on pasture. Do we need another system and will it "hold water"? So far, we have the pasture pen model popularized by Joel Salatin, and the chicken tractor model popularized by myself and Patricia Foreman. Then there is the American Free Range system that Herman Beck-Chenoweth promotes.

Do we need another system? Well, maybe you don't if you are pleased with your results. However, over the past several years of testing and retesting these systems, some of us feel the new Day Range, that I'm about to introduce, will better serve your needs.

Day Range is a model that mimics natural tendencies and optimizes performance without compromising the health, welfare, or safety of the birds. It regenerates the soil and enhances the value of the ecological systems that support both animals and humans.

If I had to sum it all up now, I'd say the Day Range system enables you to grow bigger flocks, with less labor, fewer death losses, and more profits. If all that is true, and if this model "holds water", then it is certainly worth learning more about, don't you agree?

One of the challenges I'm faced with is, that while I'm explaining why the Day Range system works so well, I am also pointing out reasons why the other systems don't work so well. Sometimes this leads to misunderstandings. Some growers who have been using a certain style of pen or technique feel challenged when I point out what I consider to be the shortcomings of that system. I am not attacking the user of the system, but rather identifying what I think are serious design concerns in the techniques they have been taught to use.

And rightly enough I accept my share of responsibility for having taught folks about chicken tractors and pasture pens. I now realize the chicken tractors can be improved, especially for larger flocks and commercial production. Consequently I have modified and changed my operations so that it "holds water better."

Day Range is an integrated system for raising broilers, layers, and turkeys on pasture. It uses sturdy portable shelters that have floors with deep bedding. These shelters protect the poultry from weather, literally from their tops to their bottoms.

The other key factor in Day Range is portable electric poultry netting. During the day, the fencing steers the poultry to areas where we want them to forage and fertilize the soil. At night the birds go back in their shelters.

The Day Range system borrows heavily on what was common practice in the early 20th century before confinement poultry housing became the norm. Much of the pasture-poultry information we now rely on comes from research work done by the USDA between 1920 and 1955. We are trying to update and re-verify some of this research, particularly the indicators for weight gain conversions and pasture versus confinement grain consumption patterns.

Day Range is the name we have given to this particular system of raising poultry that is closely modeled after other successful animal tractor systems such as chicken and pig tractors. The idea originally came from Bill Mollison, the founder of the permaculture movement, which began in Australia back in the 70's.

We have researched and embellished Bill's original ideas and ideals in our book *Chicken Tractor: The Permaculture Guide to Happy Hens and Healthy Soils*. Chicken tractors are good for small flocks of 3 to 50 hens as you might have in a garden or homestead. The Day Range model can be used on a larger production scale for hundreds or even thousands of birds.

We have used many principles of permaculture to design the Day Range system. The Day Range "system" combines free-ranging poultry, rotational intensive grazing, and efficient human management in harmonious and beneficial ways. The multiple goals of Day Ranging are:

1. Protecting the poultry from weather and predators.

2. Utilizing pasture forage, converting it to food that can be consumed by humans.

3. Fertilizing the pasture and regenerating soils, as well as collecting the bedding in the shelters to compost and fertilize our gardens.

4. Raising the best-tasting poultry in the most humane way possible.

5. Giving specific attention to the bird's quality-of-life, and their natural needs by optimizing, rather than maximizing, stocking rates. This relieves the stress associated with overstocking.

6. Providing shelter and protection from the elements. This includes giving access to sunlight, shade, and ventilation as well as protection from wind, rain, heat, cold, and wet ground.

7. Providing clean pasture and forage rotations, clean and plentiful water, supplemental diet of grains (organic when available) and minerals, and safety from inclement weather and predators.

To visualize the above multiple goals the metaphor of a tractor is helpful to describe the Day Range system. We use the perim-

eter fence as the brake to keep poultry from straying too far and to keep out predators. We use the portable shelter, water fountain, and range feeder as the steering wheel, accelerator, and driver's seat because the poultry goes where their feed, water, and shelter are.

With careful observation, we move the poultry across the landscape in a manner best designed to mimic their natural traits. We use you, the farmer, as the ignition; you stop and start the system in the way that is most productive and most beneficial to the poultry, and most convenient to you. If your birds are happy, you will be happy too, because they are putting on weight, fertilizing your fields, and will soon grace a thankful customer's dining table.

In the Beginning

Years ago, many of us homesteaders and cottage farmers (and some contrary farmers) were eagerly exploring the idea of using portable pasture pens and chicken tractors to raise poultry on pasture. Along the way though, we have also uncovered enough not-so-successful stories to give us pause as to whether to continue pulling pens or develop a better way.

This whole Day Range thing started innocently enough, when Patricia came home from a flea market with 5 little turkey poults. That was almost a decade ago, and in hindsight, it was a life-changing event. We had such fun raising the little gobblers that we bought 50 more to start a small-scale farm enterprise that eventually grew to over 500 turkeys, 400 laying hens, and 2,000 broilers per year.

In the beginning, we were raising our broilers and layers in "chicken tractors," which were small, portable, bottomless pens that we moved around our garden periodically. These were okay for small-scale summer use, but when we added turkeys to our farm, we learned right away that turkeys quickly grew too big to fit in the chicken tractors. That caused us to retrofit all our "chicken tractors" with doors to let them out into pasture. You might say it was a great leap into the pasture. Actually, it was a leap of faith for us, and a leap of freedom for our birds.

Now, we are looking over the fence into the future to see how pastured poultry systems will work. From our observations, we expect less labor, bigger flocks, and better returns. For starters, we have a new model for broiler, layer, and turkey production that merges the best tenants of pasture pens with the best characteristics of free range. We call this the Day Range model.

We believe the future of pasture poultry will be in systems that require less work, can handle more birds humanely, and will ultimately provide a greater return on investment.

Adaptations and new ideas still flow like the seasons across the landscape. New pen designs, more complete feed rations, better marketing, and a host of success stories tell us we are definitely on the right track. In short, the system "holds water." The ideal Day Range system has yet to be developed, there is always room for improvement. We leave that to the poultry pioneers of the future.

Chapter 1: Why Raise Poultry on Pasture?

You don't need a lot of money or time to start a reasonably profitable part-time business growing poultry on pasture, but you do need a willingness to learn what is required to provide the birds with sufficient shelter and food. And, you will need to learn how to process and market the birds profitably.

These three things: production, processing, and marketing are the focus of this book. These are actually three separate jobs and even professions, which makes the producer's cooperative discussed later very attractive for those who simply want to produce. However, an individual farmer can do all three jobs on a small scale and be successful. This book will get you sufficiently prepared to begin, wherever you are, to create your own successful and profitable poultry enterprise.

Even if you live in an apartment in the city, you can still begin a poultry business. Here's an example of how that can work. We have one customer who came to us with a plan. She wanted to buy large quantities of processed broilers from us at wholesale prices, and then resell them to her friends and neighbors at retail. Her hope was that she would make enough money to supply her own family for free. And, at the same time, she is developing a customer list so that she will have ready, willing, and able buyers for her own poultry products when she starts her own Day Range poultry business some time in the future.

What is in this for you? Here are the eight advantages of growing broilers, layers, and turkeys on pasture:

1. Excellent return on small investment.

2. Additional farm product that is easy to sell, and compliments other farm products such as vegetables, fruits, or meats.

3. Exceptional product for direct marketing.

4. Family friendly — pleasing to visitors and customers.

5. Will work in an area too small for conventional live-stock.

6. Little time required.

7. You can store or freeze the crop for later sales.

8. Pasture use and renovation.

Let's examine each of the above advantages in detail.

1. *Excellent return on small investment.* I can't think of an easier farm enterprise that makes a faster return on investment. Buying the chicks, brooding and growing out, processing and delivering to customers - all this with a 2-month turn around time for broilers, and four months for turkeys. Layers need five months to mature before they begin laying eggs for you to sell.

The most expensive part of your initial set up will be the processing equipment. To get started, a table top plucker and scalder can be purchased new for less than $1,000. This will be sufficiently fast to make processing a small number of birds quick and efficient. You can purchase more automated systems that will enable you and a crew of three to process 100 to 300 birds a day for under $15,000.

If you plan to start out by growing 100 or so broilers at a time, you can get by with a small floor area for brooding, and a small, portable pasture shelter and one electric poultry netting and charger. All of this will cost less than $500 and will last for many years with proper care.

Your largest ongoing expense will be feed. You can buy it already bagged from the local feed store, or arrange to have it custom mixed at a local mill and delivered in bulk. We found bulk feed was much cheaper than bagged and easier to handle. We calculated that the money we saved from not using bagged feed paid for our 4-ton feed storage bin in about a year.

2. Additional farm product that is easy to sell. If you already have farm customers buying other products from you, it will be easy to introduce your new poultry products. If you are just starting out in developing your farm business you will find that some of the easiest farm products to sell are broilers, turkeys, and eggs. In most areas of the country you will find a ready market for your products. Remember, it seems the more rural your locality, the lower your selling price might be. If you are near a large city with an upscale population you will be able to sell your products at a much better price, thereby ensuring a more profitable operation.

3. Exceptional product for direct marketing. A pasture based poultry business only makes sense if you can sell your products directly to consumers or to wholesale outlets that are willing to pay you a premium price. If you try to compete with commercial poultry prices you will not make a profit. Supermarket poultry prices are simply too low to cover your added costs of small-scale pasture-based production and on-farm processing.

There are increasingly more and more television and newspaper stories about the unhealthy conditions of commercial poultry production and unsanitary processing facilities. These can cause food-borne diseases. The good news is that more people are searching for food producers such as yourself to provide good food that is raised right. They are ready to buy from a farmer they can trust. They are willing to purchase home-grown products that are less convenient to buy than just going to the grocery store. They also tend to have the income needed to pay the higher prices that your premium custom products deserve.

4. Family friendly — pleasing to visitors and customers. When visitors come to your farm you will be able to let them walk around and view your livestock and gardens. Small scale diversified farms are usually aesthetically pleasing to the eye and exciting to the "wannabe" farmer in most folks. Many city folk long to be in the country on a homestead or farm such as yours.

One of the most surefire ways to get your farm recognized is to host school tours and to hold frequent "open farm" days for visitors to learn about your farm. The kids have a peak experience and the parents are happy they can buy good food. Some farms charge a tour fee, such at $5 per child. For a bus load of kids, that could be a worthwhile sideline.

5. *Works in areas too small for conventional livestock.* Robert Rodale believed that many of the home gardens of today will be the mini-farms of the future. This new generation of small, intensive farms will fill in the chinks of land left over from suburban and urban development. The large farms today are not suited to produce the fresh, grown-in-living-soil foods that are the most nutritious and tasty. The trend in agribusiness is towards producing semi-synthetic foods from bulk, genetically modified commodities, like soybeans and corn.

Many homestead garden-farms are very small – too small for conventional farming such as a cow-calf operation to be successful. You might have only one acre to use and your primary interest might be market gardening. You can grow crops on 1/2 acre and poultry on the other 1/2 acre. Next year just switch sides and grow vegetables where the poultry used to be. Let this year's poultry clean up the old garden space and lay down fertility for next year's market garden. You can keep this rotation going year after year and with each rotation the land will get better and better and your skills will improve to a point where your profitability increases dramatically, even more than if you were to expand onto more land.

We have heard of people raising chickens in very small places, such as in their garage, in a small back yard in the city and even in a tiny backyard in a mobile home park. As long as you keep the operation clean, odor free, and noise free, you will meet very little resistance from your neighbors. This is especially true if you introduce your neighbors to the fine full flavored taste of the eggs and meat you are able to sell to them.

6. *Little time required.* It takes nearly the same amount of time to care for a few hens as it does to care for a batch of a 100 broilers. These daily chores dovetail quite nicely with other farm work or even a full time off-farm job, and it is easy to expand when the time comes to increase your income.

At our farm, we planned on two people spending about two hours daily to do the routine chores necessary for caring for 2,000 broilers, 400 layers, 500 turkeys, 200 breeder hens, and the associated hatchery. If we were only doing broilers or turkeys, we could cut that time considerably. We suggest you keep your day job, especially in the beginning while you are building up your business. Use your evening and weekend hours to do the major tasks associated with keeping poultry.

7. *You can store the crop for later sales.* If you process more broilers or turkeys than you can sell right away you can freeze the extras and in many cases you can keep live birds for several weeks until a market becomes available.

Here's an example: we usually have a few turkeys left over at Thanksgiving. We sell most of those at Christmas, but any that remain are simply turned into turkey sausage, ground turkey, turkey breast, and turkey leg quarters. These take up less freezer space and we sell this meat throughout the year to customers who come to our farm for other products. We find these value-added products add significantly to our farm's profitability.

8. *Pasture use and renovation.* Soil "uppens" instead of deepens. You have to build up your soil, not plow it down. Today's farms often have very poor soil. If the pasture has not been maintained, the

forages will be low in nutritional and mineral value. The land will also be full of weeds and undesirable species. However, poultry can still use some of these plants as sources of vitamins, minerals, and enzymes. Birds are not cud chewers. They don't chew grass. They are not able to digest most of the grass they do eat as a feed, but can use it as a supplement. The greater good of having the poultry on pasture comes from sunshine, exercise, cleanliness, and fresh air.

Grass-based poultry is a warm and fuzzy idea and we are quite fond of it. Intuitively, we feel that having the poultry on pasture is a good thing because they will get a portion of their diet from it. Unfortunately, our research shows that just the opposite is true. We have found that it takes up to 50% more grain feed ration to grow the pasture-raised birds to market weight as it does in a confinement system.

If left to their own, poultry will graze off all the succulents, leaving behind only the coarse grasses and weeds they don't like. As your forage improves you will want to add new seed to the pasture at regular intervals.

Why People Pay a Premium for Home Grown-Food

1. They buy for the taste, flavor, texture, and the sense they are buying and eating something that has been raised well and is good for them.

2. In the case of turkeys, during the Thanksgiving holiday, folks are making only one or two big purchases, and price is less of an obstacle for that special family gathering.

3. Home-grown food is free of hormones and antibiotics. Whether it's true or not, many of today's consumers fear that the food they are eating may be laced with chemicals, antibiotics, and hormones, some of which may be carcinogenic. Actually, there are no growth hormones approved for use in poultry, and those that have been tried often result in weight

losses instead of gains. Antibiotics are used only sparingly, and not at all in organic and Day Range flocks. The biggest problem lies with the feed ration given to commercial poultry because it often contains meat by-products and grains containing fungus and molds. Consumers will go out of their way and spend extra money to buy food they perceive to be clean and wholesome for their family.

4. Many of today's aware consumers sense that they need to support local farms to keep them in business. This not only helps support local food self-sufficiency, but it also preserves open space and view corridors across farmland. When farms go out of business, the landowner is often forced to sell to developers who pack the land with new houses. Where will it all end? Without farms, where will we get our food? More people means fewer farms. Something doesn't add up here.

Zero Mortality: A Compelling Design

Usually, designing a new farm enterprise has a lot to do with how much money, time, and market share we have available. But what about the animals? Rarely do we look at systems design through the eyes of our livestock. Raising poultry on pasture gives us a premium custom product that commands a premium price. Simply put, our task is to satisfy our customers.

However, if we broaden our scope of design to include quality-of-life issues for both us and our livestock, then our management system changes. We begin to examine the environment where we force our poultry to live. If you look through their eyes, you will get a whole new vision and understanding of how your system affects their lives and your bottom line.

For the moment, consider the poultry industry as a continuum. At the far right, we have the integrated poultry industry, where billions of broilers, turkeys, and layers are raised in huge barns with strict climate-controlled environments. On the far left of the continuum, we have broilers, layers and turkeys living in very inexpensive bottomless chicken tractors or pasture pens that provide only minimal shelter from cold winds and rain, or blistering sun.

"Which of these paradigms is best for the poultry?" is not the question. The real question demanding to be asked is, "Can't we do better?"

At the far right of our poultry continuum, a new barn to house up to 50,000 broilers will cost about $250,000 to build and equip. This figure is from a recent prospectus given to an acquaintance who is considering building a broiler barn on her farm. That 1/4-million dollar investment pays for zoning and building permits, site engineering, building design, land clearing and grading, building construction, automatic ventilating and heating systems, feed bins, automatic feeders and water fountains, and training for the new owner-operator. After all of that, the owner-operator can expect to raise 5 batches of broilers per year, with 40,000 broilers per batch. For working full time, raising 200,000 broilers per year, the owner-operator can expect to earn a pre-tax, net income of about $20,000.

In almost all cases, the 1/4-million dollars to build a new broiler barn is borrowed from Farm Credit Services. The land and buildings serve as collateral for the loans. In most cases, the building itself will not be paid off before it becomes obsolete. I recently spoke to a banker who handles these kinds of loans, and he feels that none of the growers he works with has been able to earn the amount needed to pay off the building and make a decent living.

It is my contention that if any of these farmers had that $250,000 available in cash, they would not likely invest it in a poultry barn for an expected annual return that is usually less than the interest rate on the mortgage. They would more likely invest it in a mutual fund that would return about 12%, or at least $30,000 passive income, without having to lift one finger or do any work.

What this really means is that farmers are borrowing these large amounts of capital in the hopes of creating a good paying job for themselves. Instead, what we are beginning to see in the industry is offshore production and migrant labor filling the gap. I think if the commercial broiler industry ever collapses it will not be because of consumer complaints or from government interference. It will be because the rate of return has crept so low that business-savvy farmers are no longer attracted to the investment.

But what about zero mortality and the view at poultry eye level? After the farmer gambles her land, and commits all her working capital and assets to a commercial broiler house, are the poultry actually raised in what we would consider humane conditions? No, not really. To pay for such an expensive facility every single square foot must be used to maximize productivity. Broilers are crowded in at 3/4-square foot per bird. There is no fresh air or sunshine. In fact, the broilers never see the light of day until they are loaded onto trucks and hauled to the processing facility.

Bedding for large scale commercial facilities gets expensive, so they scrimp where they can. Because there is often inadequate bedding to buffer the manure and inadequate mechanical ventilation, the ammonia builds to unbearable levels, burning the eyes and throats of the poultry and requiring the farmers to wear oxygen-breathing devices when they work in the barns.

In this big industrial poultry scheme "mortality" is only an expense because it costs money to lose birds. No thought is given to the comfort of the poultry or to inhumane conditions or handling. As long as the birds grow to a reasonable size on a minimal amount of feed within an acceptable period, the grower and the parent company are satisfied.

Charles Wampler, Jr. is the son of the man who started the integrated poultry business in Central Virginia some 50 years ago. Wampler, Jr. has been quoted as saying that when he was young, it took 16 weeks and 20 pounds of feed to grow an acceptable broiler. Today, it takes 7 pounds of feed and 7 weeks to get the job done. To agri-business, this looks like progress. But to those of us in the know, it's just so much chicken manure.

In this negotiated culture in which we live, consumers blithely go from day to day marveling at the cheap and plentiful food supply, seemingly oblivious to the resulting degradation of our environment, and the foul odor that often taints commercially produced poultry. Only a very few ever raise a voice against the way poultry is grown, and even fewer still choose to do anything about it other than mutter condemnations.

But, those of us who have chosen to do something might have swung the pendulum too far in the other extreme. I am not defending the confinement poultry industry. However, in our haste to distance ourselves from the integrators, we should not ignore the very real benefits of raising poultry in confinement. For one thing, confinement-raised poultry are inside an insulated building. They rarely get too hot or too cold, and they are never subjected to scorching sun or cold rains.

In our erstwhile pursuit of better designs for outdoor poultry production, we've either overlooked our mortality rate or chosen to ignore it. It is sad that some pasture poultry farms routinely lose 10% or more of every batch of broilers or turkeys they start. In extreme cases, growers have lost 50% of their turkey poults in one rainstorm, or 75% of their layer chicks in an overheated greenhouse brooder. In how many businesses is it possible to kill off 1 to 5 out of every 10 products you raise and still make a profit?

In our years of growing broilers in chicken tractors, we have lost poultry to cold rain, cold nights, predators, and heat stress. The question we have failed to ask ourselves until just recently is, "why put the poultry out there in the first place if so many will suffer and even die?" We always assumed that because we were growing animals "on pasture" and not "in confinement" then "our" model was better than "their" model.

The reality is, we pasture poultry producers lose far too many broiler chicks and turkey poults because we put them in circumstances where they simply cannot survive. For years we've railed against genetics as the primary cause of mortalities. But what we've chosen to ignore is that the broiler strains we are using are pretty darned tough. After all, they can survive the way we treat them. If you overheat a chick one day and then drench it with cold water and chill it to the bone the next day, and it dies, can we blame that on genetics? Is this model any more humane than "their" model?

When a broiler chicken dies it might be sudden, as in the case of being crushed during a pasture pen move. Or the death may take weeks, and finally occur at the end of a long period of stress. Only a few chicks may actually die from heat stress under a baking August sun, or hypothermia during a cold rainy night in May. What we overlook is that ALL the chicks in that flock suffered the same stress, but only the weaker ones died. The survivors will not perform up to their potential. And we blame that on "genetics"?

We've heard some growers say that these modern day broiler chicks and turkey poults are predisposed to die for no apparent reason. In some strains of Cornish Crosses, that might actually be true. They seem to have the apparent genetic potential of suicide by over eating.

However, the other Cornish Cross broilers strains such as Cobb, Arbor Acres, Ross, Peterson, Hubbard Classic, and Hubbard White Mountain gain weight quickly, efficiently, and without leg problems, but, only if they are given an adequate diet and plenty of clean water, warm and dry conditions, and protection from crushing and predators.

If some of this is redundant, it is out of frustration. After years of needless losses in our flocks, we have changed our primary design. We no longer make futile attempts against all odds to make the poultry go on pasture straight from the brooder. Taking young chicks from an 80-degree brooder that is protected from wind and rain, and putting them on cold, wet grass in 50-degree weather just to get bragging rights to "grass fed" is foolish.

The jury is still out on this one, but it looks to us like the grass has very little value to the broilers anyway, other than a minimal amount of vitamins. Certainly they can't gain weight on grass, even on the lush spring growth, and especially not in the dry summer. At our farm we typically use 16 pounds of feed (more than twice as much as "their" model) to get a 4 pound dressed-weight bird, and it takes us a week longer to do it.

So, why do we want to double our feed bill and kill off a bunch of chicks just to be able to say they were grown on pasture? Surely, we can take the time to harden them off from the brooder and give them their last 2 to 4 weeks on grass, which should be enough to produce a healthy bird with a free-range flavor. Asking for more than that just causes more feed consumption and unnecessary mortalities.

What we are advocating is to take the very best aspects of confinement models and merge them with the very best aspects of pasture-based systems. The resulting model will be one that provides the poultry and the grower excellent environmental and biological security, a high quality-of-life, a low investment and it will optimize — versus maximize — the pasture-based economic profile.

In this day and age when we can literally do business at the speed of thought, there is no reason to cling to the pasture-based poultry models of the past. When we say past, we mean anything older than yesterday. That's how fast our world and everything in it is changing.

We have at our fingertips the Internet. More specifically for us, we have access to like-minded folks through discussion groups. One group I started and moderate is called: dayrange-poultry@ www.yahoogroups.com. This forum is made up of growers from every part of the country and is the biggest and best informal design team ever assembled for pasture poultry. List members range from just starting out with a few hundred birds to those who have many years of experience with many thousands of birds. When newcomers and old-timers alike ask questions they stimulate dialog between growers. Often enough there is an argument, but the outcome is always that we learn more than we knew before.

We can use the Internet as a design evolution tool where we can instantly exchange information and experiences. We can compare on-farm research results and debate the merits of a hundred different ways of doing the multitudes of tasks associated with a successful, profitable, and satisfying pasture poultry business. Instead of going through an entire season to test our suppositions, we can read the archives, or query the list and get instant feedback from one or several growers who have already tried whatever it is we're thinking about. Even if they haven't tried it, they can offer candid opinions of the idea.

There are several ways we think we can come closer to zero mortality in our broiler flocks. One is through better chicks.

Not by new genetics necessarily, but by maintaining on-farm (or regional) breeder flocks. When we order broiler chicks from any of the national hatcheries, we can never be sure which of the several different Cornish Cross strains they are sending us. What worked well on the last batch may not work at all in future batches.

Several of us are raising broiler chicks from hens raised on pasture, either from our own breeder flocks or from breeder Tim Shell in Virginia. We don't have research results to back this up, yet, but our collective experience has been that the pasture peeps show better health, more aggressive foraging, and faster weight gains.

Not every area of the country has a regional pasture-based breeder flock, so you may need to start your own on-farm hatchery. It's not hard or expensive to do this. Low cost incubators are readily available, and keeping a breeder flock is no more difficult than keeping a layer flock. An on-farm hatchery gives you chicks that are not stressed in transport. Taking them directly from the incubator's hatching tray to the brooder means we don't have to rely on airlines that won't carry them except during certain temperature ranges, and post offices that leave them sitting on a loading dock during a cold, windy, rainy night.

Better feed can also have a positive effect on broiler and turkey performance. We've always considered that the ideal small farm system would be one in which the poultry manure is used to fertilize the fields where the poultry feed is grown. It takes about 3 acres of grain field to produce the feed for 1,000 broilers. Interestingly, the manure from 1,000 broilers is enough to fertilize about 3 acres of land.

We don't have research results to prove this, but our intuition and common sense tells us that grains grown organically, on a healthy, well-balanced soil, will grow better poultry than grains from a depleted, chemically-dependent soil of the type that predominates in our national landscape.

Better shelter is another major key to survival in pasture-based poultry systems. We are not defending the confinement broiler houses with their environmental controls. These facilities are simply too expensive, and don't offer the access to the outdoors and green growing pasture that we want.

Likewise, we are unable to defend the typical 10- x 12-foot pasture pen or the 4- x 8-foot chicken tractor. These pens are too labor intensive for the small number of birds they hold. More importantly, they simply do not provide the protection from cold wind and rain or blistering sun that these birds need.

Some of us are working with the Day Range system that uses mini-barns with floors that are on skids. This is combined with electrified poultry netting to shelter the broilers, layers, or turkeys, and to protect them from predators. This system works remarkably well, given that it can also be up to 50% less expensive and up to 75% less labor intensive than other pasture poultry models.

There are some drawbacks, of course. One is that the system requires more thought from the human manager. Rather than just routinely moving the pasture pen each morning, the manager needs to be able to read the grass and understand the poultry behavior to tell when it is time to move them to a new site. Another drawback is the airborne predators. The electric poultry netting will stop all 4 legged predators, but we don't yet have a good way of locking out the hawks and owls.

We can usually prevent losses to owls, which are nocturnal hunters, by securing our birds inside the shelter each evening. This has the added benefits of protecting the birds from cold night rains, and of capturing up to 1/2 the manure in the bedding so we can compost it for our market gardens.

Pasture pens and chicken tractors have brought us this far, and they are "holding water" good enough to keep us going. Let's continue to use them, hopefully more wisely than we may have in the past, until such time as a new and demonstrably superior system presents itself.

FREE Ranging or DAY Ranging?

Funny things sometimes happen to our favorite expressions in the English language. For instance, the words "all natural corn flakes" conjures in our minds an image of wholesome food grown by family farmers and minimally processed by local companies. Instead, we get genetically-engineered corn grown on soil depleted farms in the Midwest. The corn is sent to the east coast for intensive processing, and packaged in a box that used to be a tree in the northwest or southeast, After all that, it is put onto a tractor trailer or freight car for a ride to the distribution center for further shipping to a store near you. How "natural" is that?

It's the same thing with free range poultry. Fifty or sixty years ago, back when farmers were still family, and Madison Avenue hadn't yet invented "all natural", a free range chicken or turkey probably had been grown on pasture. It was most likely raised on a small, family-owned farm where it was also processed and sold to a neighbor or to the local general store. In short, free range meant ALL NATURAL, in the best sorts of ways.

Then USDA got into the picture, and decided it would be just fine to call the product free range as long as the poultry had access to the out-of-doors. What that means is, that in a 40,000 unit broiler barn, you can leave the barn door open and magically become a "free range farm". I prefer to think of this as just an oversight on the part of the USDA, and not a concentrated effort to dupe the buying public.

Unfortunately, the buying public is deceived, and it's no wonder that public perception of agribusiness and the USDA is one of wariness approaching disbelief. We don't know how many times in the past ten years we've heard customers exclaim how good our homegrown chickens taste compared to the store bought chicken they used to eat. Small wonder when you think about it.

At some point, as the Day Range community of growers broadens it base, we will want to approach the USDA about their obvious oversight in confusing terminology. The reality of confinement-based poultry production is such that using the term " free-range" is indefensible.

Lately, we've noticed birds being labeled "free roaming" and this is also misleading. This simply means that 40,000 broilers are able to "roam" freely within the 3/4 square foot space they are allotted. Surely we can do better for our feathered friends.

Chapter 2: How Day Ranging Was Started

We first learned about "animal tractors" in the spring of 1990 when we were researching ways to include livestock and poultry in the Intervale Community Farm in Burlington, Vermont where Andy was the farm manager at the time. Our hope was that we could increase soil fertility, manage and control grasses and weeds, and have new products—meat and eggs—to sell to farm members. When we learned about animal tractors in Bill Mollison's *Permaculture, A Designer's Manual* we realized that we may have found the perfect system for accomplishing our goals.

Mollison talked about the animal tractor as being a portable cage system for capturing the well-known characteristic of chickens and pigs to scratch and dig the ground in search of worms, insects, and roots. He explained how chickens, pigs, or goats enclosed in a weedy or brambly area will devour all vegetation, partly cultivate the earth, and manure the area. The farmer then rotates the animals to another enclosure to continue the process on fresh ground.

Right away Patricia and I set about building a portable enclosure and starting raising broiler chickens to test the efficacy of Mollison's claims. What we learned over the next several years and after thousands of chickens—raised by ourselves and by a host of friends—became the genesis of the book: *Chicken Tractor: The Permaculture Guide to Happy Hens and Healthy Soil.* In *Chicken Tractor,* we discuss the basic concepts of permaculture and the use of animals in food systems. Using chickens in gardens to regenerate soils was a natural progression to our book *Backyard Market Gardening: The Entrepreneur's Guide to Selling What You Grow,* which had become a flagship book for helping to establish farmer's markets and community farms across America. Our intent with the *Market Gardening* book is to help re-establish the awareness of the need for community food sufficiency, and for keeping dollars and resources local.

The release of *Chicken Tractor* coincided with a renewed national interest in good tasting, humanely raised food. Thousands of people across the country followed the instructions to build a "chicken tractor"and began raising small flocks of chickens on their homesteads or in their backyards and gardens.

We can describe a animal tractor as "an integrated system for raising poultry or livestock in portable shelters on pasture." This model mimics natural tendencies and optimizes performance without compromising the health, welfare, or safety of the animals. It regenerates the soil and enhances the value of the ecological system that supports both the animals and their human managers.

Our original intent was to speak of the "chicken tractor" as a holistic rotational system, with the pen being only one part of that system. As time went by, however, the term "chicken tractor" came to mean the actual pen itself. Ideally the pen is moved daily, or as needed, to give the poultry fresh graze and to spread their manure more uniformly over a larger area, thus creating an integrated system.

For smaller backyard flocks (3 to 100 birds), the portable pens are semi-satisfactory in good weather. The chicken tractor or pasture pen is usually a 4-foot by 8-foot or 10-foot by 12-foot structure, that is 3-feet high. The top is hinged as a lid for access, and the water fountain and feeder are located inside. In the case of egg layers, the shelter pen usually includes a roost and a nest box. Anywhere from 10 to 30 hens or 20 to 80 broilers can easily be kept in a pen of this nature, depending on the actual size.

A chicken tractor pen is small enough that it can be moved as needed, especially when on wheels or by using a wheeled dolly. The covered lid protects the chickens from overhead predators and from sun and rain. The poultry wire sides prevent predators and dogs from getting at the chickens, and the frequent moves keep predators from digging under the sides. The chickens are directly on the ground with easy access to weeds, grasses, seeds, bugs, and grit. All of these circumstances—combined with a well-developed, natural grain ration—result in wonderfully flavored broilers and eggs that are easy to sell for a premium price.

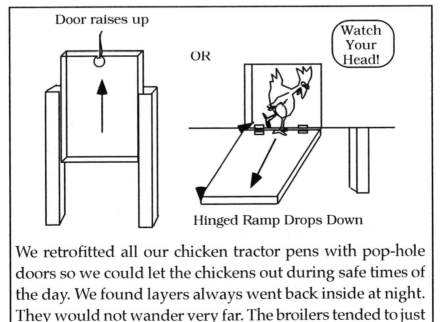

We retrofitted all our chicken tractor pens with pop-hole doors so we could let the chickens out during safe times of the day. We found layers always went back inside at night. They would not wander very far. The broilers tended to just bed down wherever they happen to be when night fell.

When it comes to raising a larger number of birds in this size pen, however, you encounter problems of space and height. This is especially true of turkeys. A full-sized turkey tom has a wing span of about 6 feet and can stretch to a height of 4 feet or more. Even if we only keep a few turkeys in this type of chicken tractor we feel they are unable to express themselves naturally, and that their environment is not the most humane we can provide.

Another concern is to find ways for pasture-based poultry growers to handle larger numbers of poultry in order to earn a significant income. We began to research this question and to talk to people who shared similar interests. Here's a little of what we found.

Free-Range Turkey History

Up until about 1955, all turkeys in America were raised on pasture in a "free-range" model. We found a wealth of information in out-of-print textbooks. From the 1920's until about 1950, the United States Department of Agriculture had placed a tremendous amount of emphasis and research on raising free-range poultry.

Their research was intense. For several years, researchers at the USDA Greenbelt facility worked with Charles Wampler (progenitor of the Wampler-Long Acres brand) to develop a grain ration suitable for each type of pasture and time of year. If the grower had alfalfa and bromegrass pasture, then he would use ration #31, for example. If the pasture was fescue, he would use ration #38, and so on. The rations were calculated to include the amount of protein available from each type of graze.

Our search for information was aided a great deal when we discovered that our next door neighbors, twin brothers Russell and Ronald Fleshman, grew up on their dad's free-range turkey farm which at that time actually encompassed the 40 acres where our farm is today. They shared wonderful old photographs with us, showing them as young boys carrying buckets of feed to the turkeys. Because they were such small kids at the time, they carried sticks to keep the turkeys from pecking them.

Coincidently, the Shenandoah Valley where we live was at one time the leading producer of free-range turkeys in the country. The area is rich with history and information about methods to raise free-range turkeys, and all we had to do was ferret it out. Rockingham County, near the center of the Shenandoah Valley, is still the turkey capital of the world. Millions of turkeys are raised in confinement in the county each year – more than in any other county on Earth.

Twin brothers Ronald and Russell Fleshman working with their dad's range-raised turkey flock. When you are about the same size as a turkey, you need to walk softly and carry a big stick. Photo taken circa 1950s.

One of the things we learned from the past was that the free-range industry had a very low potential profit because of high labor costs. It required a tremendous number of person hours for flock management; six people were needed to run a 10,000 bird flock. Fencing was always a problem. They didn't have our modern day electric fencing technology. On top of that, the turkeys of the time were lighter and wilder, and they could fly easily. The free-range industry was also fraught with potential disaster from predators, parasites, and diseases—especially blackhead, a deadly turkey disease.

By the late 1940's and early 1950's, the industry was moving more and more towards confinement raising. The beginning of this movement was to put the turkeys in buildings with attached wire-floored "sun porches". These units were elevated about 3 feet so the manure could be raked out from underneath.

This dramatically reduced the number of person hours required for feeding and tending the flock. Fencing and range management were eliminated, and feeding and watering chores were concentrated into a much smaller area. More importantly, these elevated buildings protected the turkeys from predators and kept them up off the ground to prevent transmission of blackhead disease.

However, a new set of problems cropped up however. The new confinement houses were expensive to build, and when confined into too small an area, turkeys and chickens proved to be prone to cannibalism. The wire floors cause leg and foot problems as well. Also, eliminating the pasture component limited their exercise and muscle toning, and the poultry become more vulnerable to sickness in this confined model.

While on the one hand confinement growers had reduced their labor costs for range management, they had increased their production costs due to these other factors. In the end they were probably worse off than when they started.

By the mid 1950's, USDA researchers had turned their full attention to researching and teaching methods to raise turkeys and chickens in total confinement. Huge barns became the new model, with some barns today covering several acres and costing upwards of a quarter of a million dollars.

Wonderfully efficient, it is now possible for one or two people to raise 10,000 or even 20,000 turkeys or 50,000 broilers at a time. However, is it profitable? If you add in the cost of running ventilation fans, feed conveyors, water lines, trucking in bedding, trucking out manure, the profit margin drops. Next, include the cost of overcrowded animals which can lead to cannibalism and poor flock health. Now add the low price per pound paid to the growers. The end result is thousands of birds, hard work, an expensive environment, and lower income.

Most importantly, in this equation we ignore the welfare of our poultry. There is a parallel mind set with how medicine is practiced in America; our commercial health care system (which is really disease management) tends to treat symptoms rather than causes. The commercial poultry growers also tend to react to problems with short-term Band-Aid solutions that fail to address the underlying problems.

For example, a serious problem with confinement is that birds peck and even cannibalize each other. Solution: clip their beaks (upper and lower) and in the case of tom turkeys – cut off their snood. I've noticed that folks in crowded cities seem to peck at each other, and even kill each other. Should we clip their upper lips and noses? No, the root of the problem is simply that the birds are too crowded and stressed. The humane solution is to give them more space.

Problem: the birds fight. Solution: cut off their toes. If this were true for people we'd have a lot of toeless folks hobbling about. A new shoe industry would develop for stubby feet.

Problem: the old style breeds don't grow fast enough. Solution: develop new breeds and strains that will perform well under such adverse conditions.

Problem: The birds grow so fast that every smidgen of protein goes to growth, at the expense of their immune systems. Solution: antibiotics.

Problem: the birds get too inbred and become nervous or flighty. Solution: keep them calm by blocking off all light, including sunlight (which could help keep them healthier).

Another consequence of the new high priced poultry buildings is that it eliminates many, many farmers from being able to compete using their old, healthier, free-range models. As a result, the industry is concentrated in just a few companies—known as integrators—where it seems the health and welfare of the farmer is even less important than that of the poultry.

The final outcome appears to be this. We used to have tens of thousands of small to medium sized free-range poultry farmers across the nation. Now we have only a few thousand large confinement growers supplying the nation with poultry. There is a point of diminishing returns, and that point has been reached when you need to begin worrying about how toxic or safe your food supply is.

For example, meat grown in confinement facilities may be laced with all sorts of things that are not good for us to eat. An example is pesticides that are in the feed to combat insects, or arsenic in the turkey water, which combats stomach parasites. Arsenic also irritates birds' digestive systems causing them to ingest more food and thus gain more weight in a shorter period of time. Another way to reduce operating costs is to buy cheaper feed. All of this in a muddled attempt to get faster weight gains and bigger profits.

Consumers who are aware of the problems of modern agriculture are so hopelessly fed up with the bland taste and potential for carcinogens in commercial poultry that they are more than happy to travel out to our farm. They are also happy to pay premium prices for our custom grown products.

We need to produce what customers want in numbers that will satisfy this expanding demand. And, we want to do it with a system that has the permaculture goals of building soil and integrating our gardens and pastures with livestock for even better future production. All this is possible with the Day Range system.

Chapter 3: Day Range Poultry as a Business

Benjamin Franklin said, *"The way to wealth, if you desire it, is as plain as the way to market. It depends chiefly on two words: industry and frugality. That is, waste neither time nor money, but make the best use of both."*

It's as true today as it was in the 18th century — we can save money, but we can't stop time. We must strive in every way possible to make our poultry business efficient <u>and</u> effective. Make every move count, then count every move so that the least effort will have the greatest reward.

Farmers who lack investment capital mainly have time and energy as assets. We are among the most proficient workers on earth, yet we often fail to account for the time we "spend" in our business. Typically, farmers will pay everybody else first, then be satisfied with what is left over. We knowe this is true, because we managed our time and business that way for many years. Finally we realized that our farm business wasn't profitable enough, so we changed our operations. If we value our time at "fair market value", perhaps we won't be so apt to fritter it away.

If a semiskilled worker in an automobile assembly plant gets paid $20 per hour, then a semiskilled farmer should expect the same — or better. Yes, we know all about the "side benefits" of farming: fresh air, sunshine, self-direction and self-employment, and on and on. What that doesn't buy you though, is a decent pay check. Honestly, if you find yourself arguing with us on this point, then you are very much a practitioner of the "poor farmer" mentality.

On one of the e-mail discussion groups we subscribe to, a woman was arguing fiercely for her right to earn $6 per hour baking bread to sell at farmer's market. This from a woman who obviously has certain life-skills that should enable her to make many times that much per hour. Yet here she is arguing that $6 per hour is "good pay for a farmer".

Farmers are worth more than minimum wage!

How sad that farmers who provide food for others feel they are only entitled to work for minimum wage. What's even sadder is when these "underpaid-by-choice" farmers try to impose such limitations on the rest of us.

Owning a farm (even if it is mortgaged) plus all the equipment and facilities, and the materials to raise food, is a huge investment. That total investment would earn far more per hour if we simply invested in mutual funds. There is no good reason any farmer should work for $6 per hour for their labors.

Right now in the United States, $5.45 per hour is the legal minimum wage. However, in every community in the country, the "effective" minimum wage is whatever McDonald's and Burger King are paying burger flippers. In our area that's currently $7 per hour, plus benefits. A burger flipper doesn't need any extraordinary skills or special training or equipment. Just show up, learn how to flip burgers, and earn your pay.

Comparatively, a farmer has mortgage or rent payments, costs for transportation, seed, equipment, materials, labor, and a need for a wide range of specialized skills. Please, don't sell yourself short. If you can make more money doing almost anything other than farming, then you must really explore how badly you want to be a farmer.

Some farmers are making an excellent living, while others are just getting by or even losing money each year. Very few that we are aware of are getting rich in farming. Our best return probably results from escalating real estate values. This enables the farmer to either borrow money against land to stay afloat, or to sell out some day in the future for a perceived profit.

The first directive of farm planning should be to insure the quality-of-life we have chosen. This includes vacations, time for ourselves and family, hobbies, and a chance to invest excess capital to provide for our childrens' educations and our retirement. Now, we're over 50 and will probably always be active doing something. We can't see retiring and do nothing. No second-hand life soap opera, bon bons or couch-potato life for us. But there are other things in life we want to pursue, such as travel, time to write and study, and other interests not related to farming. Those things are a big part of our quality-of-life pursuits. I'm sure it's the same for you.

How to design for your quality-of-life? Decide first what you want from your farm. Do you want a full time income, hobby farm, or a part time income? Do you simply want a quiet place to live while you work in the city?

Use your off-farm job to finance your land, yor home, and your dream, including livestock and buildings. Then, with minimal or no debt, try to make a living from it. Alan Nation, publisher of *Stockman Grass Farmer*, says a farm's operations can either pay the mortgage or pay the farmer, but usually not both.

We visited a beautiful part of America's northwest a few years ago and spent time with an old farmer who pointed out to us that it's quite common for city slickers to visit a charming rural area and fall in love with the idea of starting a farm there. Unfortunately, most of them fail after a year or so and head back to the city. His closing quote: "You can't eat the scenery," sums up how futile it is for anyone to try farming as a business only for the side benefits. Yes, we do enjoy those side benefits immensely, but we still have to earn enough money to enable us to stay on the farm.

Are You a Part-time or Full-time Poultry Grower?

Today, there are very few pasture poultry producers who are making all of their income from poultry alone. Most of the growers we know are raising a few hundred to a few thousand broilers each year, and expecting to earn only a percentage of their total farm income this way. Typically, farmers will have pasture eggs, broilers, and turkeys along with one or more farm enterprises that might include beef, dairy, pork, veal, hay, firewood, vegetables and fruits, hatchery, goats, and horses. And, almost all pasture poultry producers we know have an off-farm income.

To get to a place where it is possible to make a full time income from pasture poultry will require several thousands of broilers, turkeys, and laying hens. But don't let that stop you from beginning! Start with just a few dozen broilers or turkeys, or a hundred or so layers. Then see how you like it before deciding if you want to do this as a way of life. Making a decent profit on your investment in pasture poultry is not difficult, and you will know within a very short time if pasture poultry is a good farm enterprise for you and your area.

Passion, Burning Desire, and a Positive Mental Attitude

Begin with goals. Describe the quality-of-life you crave. Where lies your passion, your burning desire? Whatever you choose to do with your time and your resources depends entirely on what you hope to accomplish. If you don't have goals, then how will you know if you are heading in the right direction? One of the paramount goals for Patricia and me is to live the quality of life that we most enjoy. We want to live in the country, on a farm, and be able to earn some of our income from our farm enterprises. Any time we develop a farm enterprise plan and budget, we always test it to make sure that it meets our criteria to be environmentally sensitive, economically sensible, and socially just.

Some of your goal setting exercises can be enhanced if you simply ask yourself the question, "What do I want to be when I grow up, and with whom, and where?" Most who are reading this book are already grown up and have a strong inner child wanting to play on the farm — bare foot with twigs in your hair. These questions are valid, and the answers will hold tremendous power and insight for you.

Once you decide to be a pasture poultry farmer, then it's time to take inventory. List your resources, personal goals, time available, cash, on-farm and off-farm interests, skills, contentment level, and who is available to help you. Involve your family and your community in your resource inventory, so that you will have a clear understanding of their inclination and desires.

You may find, as we did when we first started, that what we really wanted was a hobby that pays a decent wage. It was only after the fact that we increased our level of contentment goal to include earning a significant amount of money from a small farm poultry enterprise. Now, we're retired from farming and are working on the sidelines to help the next generation take over the farms of tomorrow.

Let's look into the future to see what lies ahead for farmers. One way is to envision a glass half empty. We have to worry about population pressure increase, environmental degradation, saline seep, aquifer and watershed draining, a pending stock market crash, a looming farm crisis, environmental regulations, escalating fuel and transportation costs, a declining number of skilled farmers due to age and income quest, and so on. Not a very inviting picture.

Prepare for the future by expanding your knowledge.

A Golden Egg of Wisdom from Chairman Mao Chicken-tung's Little Red Book

Let's now look toward a future where the glass is half full. Thanks to innovations in technology we have expanded knowledge and improved materials. People are demanding flavor, freshness, and food safety. This opens new markets for growers of clean food. Intensive versus extensive production, enables us to grow more on less land, which helps to offset the loss of land to real estate development. It is becoming more common to look for local markets for specialized foods, versus commodity crops for national and international markets.

We are learning how to add value to our products, so we can earn more from less. We are discovering sustainable integrated systems to replace antiquated linear development of products. Education levels are increasing, not only for farmers but also consumers. This is not the typical land grant university type of education that is partly funded by the agricultural industry. But, rather the new, alternative knowledge that works on smaller parcels of land and will enable family farms to thrive. This is the atypical, "linking farm gates to dinner plates'," as my good friend Steve Bonney likes to say.

Beginning farmers are not poor, they are rich. They have enthusiasm, ability to learn, willingness to work hard, and courage to spend the time needed to succeed. Our most exciting moments lie ahead. People are seeking us out for improved flavor, freshness, and safer food, and are willing to reward us for providing them what they need. Never in our country's history has there been a better time to start a small-scale diversified farm. The time is ripe, so let's look at how we can develop farm enterprises that will satisfy the market, while sustaining the farm and the farm family.

Planning Your Farm Future

All businesses really need to begin with a business plan and a budget. Don't groan. An unwritten plan is an unmet plan. If you have reasonable estimates of costs and potential income, then you can determine if starting the new business is the course of action you want to follow. It is very clear that raising poultry on pasture, using any of the available methods, is <u>not</u> a way to get rich quick. It will be profitable if you follow the advice in this book. And, assuming you have necessary market and business skills, at some point you will be able to make a full time living doing it.

I have very little patience for people who complain that some-one is "getting too big." Too big for what? To make a living? To send our kids to college? To set aside money for retirement? Getting big is a relative term in the poultry industry, anyway. For example, all the pasture poultry growers in my state of Virginia are still producing less than 20,000 broilers and 2,000 turkeys per year combined. Compare that to the integrators in this state that are producing 1,000,000 broilers per WEEK and you can begin to see how silly it is to accuse anyone of getting "too big."

Some things you will want to include in your budget are mini-barns, fencing, chicks or poults, feed, processing equipment, marketing, land and building rental, electricity, water, medica-tions, brooder equipment, feeders and water fountains, labor, and profit. In all cases, the profit, or amount you want to make from the enterprise, needs to be entered first.

Pay yourself first. Don't do what almost all other farmers do: add up all the expenses, subtract them from the sale price, and call that profit. Sure you can do that, and you might come out okay. But I recommend you pay yourself first before anyone or anything else.

Consider compatible products that use some of the same equip-ment and techniques. For example, you can serve three market segments. 1. With broilers, by harvesting some of them young. 2. As fryers. 3. Older as roasters. At any level, they use the same feed, same processing facility, the same incubator, same customers, same marketing channels, and so forth. The only difference is time, whether you keep them short term or full term. This could also be said of quail, partridge, pheasant, ducks, and geese.

Other businesses that are compatible with poultry are beef, dairy, or pigs. The poultry can be used to sanitize the pastures after the livestock have deposited their manure. Any large animal parasite larvae deposited in the manure becomes worm sushi to the chickens and will be picked out by the poultry. This is good for the fields and high quality proteins and lipids for the birds. This is particularly true with laying hens because they like to scratch and peck. Broilers and turkeys are less interested in scratching the manure apart, but they will do some.

Pigs are particularly compatible with poultry because you can give them the same feed you are feeding your hens or broilers or turkeys. That way you only have one feed bin and one feed ration to deal with. While the protein content may not be ideally suited to pigs, it will be close enough that they can thrive on it. Pigs will catch and eat chickens and turkeys, though, so it's not practical to house them together.

How to Get Folks to Buy from You

Don't kid yourself about how easy it will be to get people to buy your food just because it's the best there's ever been. Getting customers is one of the hard parts of the home based poultry business. You are asking them to change their buying patterns when you ask them to drive out to your farm instead of to the grocery store. Because your products are only available seasonally, you're asking them to "stock up" — something Americans are not very good at any more. We're asking them to admit, at least temporarily, that the food they buy in the grocery store may not be all that good for them. We are asking them to pay us more money for food than they ever thought they would.

Never mind the fact that Americans are the best fed, least nourished people on Earth. Never mind that we spend more for tobacco, alcohol, and entertainment than we do for food. Two parents smoking and having a six-pack in the evening will spend more in a day for that than it would cost them to buy and enjoy a fine homegrown roaster.

It isn't that people don't want to buy from you; its more likely they never hear about you. We live in a county of 25,000 people and it amazes me how many folks don't read the local newspaper or watch television news. We've had numerous articles about our farm in area newspapers and have been on television news stories several times, yet almost everyone we meet for the first time says they've never heard of us. Invariably the new people we meet who have heard of us learned about us from a friend or relative. That means your sterling reputation and word-of-mouth advertising is probably the best way and maybe even the only way to attract new customers. Once you do get the customer, though it is fairly easy to get them to buy more diverse products from you.

Compatible market products are: stew hens, pet food, eggs, poults, chicks, feed, pullets, beef, pork, lamb, dairy, and so on. Once you have gone to the effort of getting a new, satisfied customer, then start adding on the products they will buy. You can increase your income exponentially by focusing on as few as fifty families in your area.

If you can get them to buy everything that you can produce in your climate zone then you can earn thousands of dollars annually by serving these core families. This is the genesis of the "50 family farm" where only 50 households provide a large enough market to reasonably support a small-scale diversified farm.

However, it will take you a while to build up your core group of satisfied customers who are willing to go out and spread the news about you. You will need to find some way of letting customers know you are in business and have really good products for sale. Running a classified advertisement is one inexpensive way to promote your products, but it still depends on readers who actually read the want ads. Several years ago when we lived in North Carolina we had good results running classified ads in our local daily paper. However, here in Virginia we have tried want-ads with very little response, not even enough to pay their cost.

Hold social functions at your farm whenever possible and show slides. Invite school groups out to visit your farm. Show them around and explain why you are doing what you do.

Some folks are so distracted by their daily lives that even if they want to buy from you they often don't, simply because they're too busy. For those folks you need to offer a way for them to buy from you conveniently, usually through a home delivery service, or by selling through a retail outlet in town such as the local health food store.

Group sales are another avenue. My friends Marlin and Christine Burkholder have a CSA (community supported agriculture) farm just north of Harrisonburg, Virginia. Every Monday evening they deliver about 50 bags of vegetables to their drop off point in the city where farm members pick up food they have ordered. About 3 times each year they poll their members to take orders for our poultry products. The Burkholders then let us know how many broilers and turkeys their group needs. We then arrange to deliver them to one of the Monday evening pickups. In this way the Burkholders generate sales and get a commission for brokering the orders, and we are the suppliers.

The places you will be most likely to sell poultry products are local specialty shops, better restaurants, health food stores, and food buyers cooperatives. Their customers are accustomed to buying clean food and paying a higher price for it. Always schmooze your own customers, too. Ask for their help. Tell them you need to find new buyers. Anytime they send you a customer be sure to reward them with a little gift of extra eggs or sausage or whatever you have on hand that you want to encourage them to buy. People are usually happy to help a small farmer, or a friendly local farm.

You can also increase sales by extending your marketing year. Buy a large chest freezer and store birds so you can sell them later in the winter. Many families don't have a home freezer. You will be doing them a favor by freezing and holding extra broilers or turkeys so they can enjoy them in the off season. We plan to have a bumper crop in November at the end of our growing season. We put the poultry in the freezer and wait patiently to let the market catch up.

Sooner or later your regular customers will come around looking for more products. This way, you can also supply your retail stores and restaurants with product throughout the year so you don't lose your place on their store shelves. By the way, we also raise our prices at least 10% in the off season to offset our costs of freezing and storage.

Make an effort to identify and contact the commercial outlets in your area that might sell hundreds of birds for you each year. You will probably need to lower your retail price and sell to these institutions at premium wholesale. A good white table cloth restaurant will use anywhere from 500 to 5,000 broilers each year. We have one restaurant that buys 18 roasters from us every Monday. That's nearly 1,000 roasters per year to just one rather small restaurant. It doesn't take many of those to make your business do well.

Another foot-in-the-door promotional idea we've used is to deliver a gift basket (or box) to each quality restaurant and bed and breakfast in our area. This gives them a chance to taste our superior custom products. This is about the only time I recommend giving away product in order to make a sale. These homegrown Day Range broilers are so good that a discerning chef will have no trouble at all recognizing their value. So, let them cook one and try it! That will help them decide to buy from you on a regular basis. Always include your farm brochure so they will know how to contact you and what you are all about.

Make it easy for them to buy from you. Send them a price and availability sheet by fax or email weekly. It also pays to call them regularly to get their order and make sure they are happy. Any time you have a new product give it to the chefs first so they can have the honor of introducing it to the community.

Farmers' markets can also be a wonderful place to sell your poultry and other products. Keep in mind, though, that there are different levels of participation in farmers markets and not all markets will let you sell meat products.

Some farmers' markets are barely holding on, particularly the ones in small rural communities where they only have a few dozen customers each day. You won't have much luck at these markets. Where you will find plenty of buyers is in the larger inner city markets. These can attract hundreds or even thousands of shoppers each day. I have friends who are selling anywhere from 100 to 500 broilers per week, and for better than average prices. Farmers' market sales don't work for everyone. But when they work well they are really worth attending.

Selling Live Poultry & Livestock

Sometimes you can do better selling your poultry live, than if you pay to grow them out, process, and store them.

Products you can sell live are:

- Chicks and poults from your on-farm hatchery. We'll talk about this more in the next chapter. Once you have an on-farm hatchery though, your products can be sold through the mail to a national market. You will be one of the very few who can provide chicks or poults from pasture raised parents. Tim Shell calls his chicks "pasture peeps".

- Pullets. Every year we get calls from families looking for a few pullets to start a small layer flock. From time to time we also get calls from small-scale commercial growers who want a starter flock. This market is fairly small, and local, but could be a reasonable outlet for some of your products.

- Second year laying hens. These are good for families wanting their own small layer flock. We'll discuss this more in the chapter on layers. We don't keep our laying hens beyond their first year of laying (18 months). Their production drops by about 20%, and they give us too many jumbo eggs. Jumbo eggs don't fit easily in egg cartons and tend to get broken when the cartons are stacked.

- Colored birds to ethnic markets. Some cultures prefer colored birds, especially during their religious holidays. We've never been able to tap into this market because we simply don't have that many ethnic groups in our county. But for those of you who do, this could become an important part of your business. In many cases they want to process the birds themselves.

- Rare breeds to poultry fanciers. This is a viable market for those of us who have an on-farm hatchery. You can hatch and sell day old chicks or poults to anyone anywhere in the country, if you ship them by mail. For more information please see the Hatchery chapter.

- Feathers. While not a big market, you will from time to time find someone who wants feathers.

This picture is of Chief, one of our broad-breasted bronze toms. We sold Chief to Bill Wilbur of Wilbraham, Massachusetts, in late 1999. Bill drove to our farm in Virginia to get this bronze bird that had opposing curls on the tips of the two middle tail feathers.

Bill makes authentic reproductions of Native American headdresses of museum quality. One particular style of headdress requires these very special turkey feathers.

This Bronze tom will grow feathers as his livelihood.

Bill also raises birds for his fly-tying market. He doesn't kill the birds. He just harvests their feathers before molting season.

You can collect feathers before scalding. We always have a basket containing feathers for folks to pick one. They really get a thrill out of having a large tail feather as a souvenir.

- Fertilized eggs. You can sell these through the mail to people who will hatch them out in their own incubators. We have tried this with turkey eggs

Bill showing some of the Native American headdresses he creates with different types of feathers.

with poor results, but we did have an 85% hatch rate for quail eggs. The jolting during shipment can harm the fragile membranes and decrease fertility.

- On-the-hoof beef, pork, lamb, or goat. We have folks stop by to get animals for their 4-H projects and breeding stock of our Boar meat goats. We get quite a thrill when one of our offspring is awarded the Grand Championship and blue ribbons in the goat classes.

- Self-Processing. There are also folks who will do the processing themselves. Sometimes you can sell the animal live, arrange for the transportation, and let your customer pick up the meat directly from the butcher. Naturally, for a fee, we will be happy to deliver the meat directly to their home.

- Other growers. You can sell birds to other growers who have more orders than they can fill. We did this for a grower just south of us. He bought 150 live turkeys. We delivered them to the processor the week before Thanksgiving. Our friend then picked them up and distributed them to his customers.

How Else to Promote Products?

Get out and explore the region and find out what others are doing. Here are a few more ideas we have tried:

- Have a BBQ at your farm during the summer and invite everyone you can think of. Not all of them will buy from you, but most of them will talk appreciatively about you to their friends. Our annual full moon, free-range festival is certainly a success.

- Barter for beef, pork, and other things to sell to your customers.

- Add value to meats by cooking or smoking cuts , or by offering breasts, sausage, etc.

- See if groups doing their annual benefit BBQ will buy from you, or even place an order a year before so you can contract grow for their event.

- Visit allergy doctors in your area and let them know you have good food for sensitive folks — the same with pharmacies.

- Brochures with tear-off tabs at various locations have been mildly successful.

- A float in annual parades will put you in front of a lot of folks. Pass out your farm brochure along the way.

- Do a web page about your farm and what products are available. Almost all of our customers have email, and we send out bi-weekly newsletters with photos, for virtually no cost.

Will Direct Mail Advertising Work for You?

Direct mail advertising for your on-farm poultry sales probably will not pay off, at least to do it more than once. In our county for example, there are 10,000 households, but only a fraction of them will be interested in premium poultry. It costs about 35-cents for each mailer, including postage and printing. It would cost us about $3,500 to send just a mailer to our county residents. If only 1% respond, that's only 35 sales. Since we only make a 25% to 50% mark up each customer would have to buy $200 to $400 just for us to break even on the advertising cost.

In the long run, however, this may not be so bad, because any one customer can be worth hundreds or even thousands of dollars each year. We must not think in terms of them buying only one or two chickens, but that they will come back to the farm numerous times throughout the year, and year after year, to buy more chickens as well as all the other products we have to sell.

I think there are better ways than mailers to access the type of customer you seek. Brochures and signs at places where food conscious people are likely to congregate is one way. For example, people worried about clean food will frequent health food stores in your area. People craving good tasting food are likely to buy good utensils, so put your brochures in the kitchen shops in your area, and at cooking classes.

We have used email advertising successfully and economically to get new customers. Check with your local provider about the marketing services they offer.

What about Competition?

Day range poultry is no different from any other farm enterprise when it comes to competition. It's a sure bet you will have some, and it's up to you whether or not they are as successful as you are. Traditionally in the pasture poultry business we have seen individual growers in large areas who don't have any neighbors who grow poultry, but that is changing. More and more growers are starting up, and even if they are only growing 1,000 or so birds per year, they still have to look for buyers in a very select segment of our population.

That means as time goes on, more small growers will be starting up and will be competing for seemingly fewer and fewer buyers. Either one grower will be more aggressive and get most of the business, or will lower prices and dramatically handicap all growers. Regardless of how you slice the pie, it is still a small pie. Each time you carve out another slice for a new grower you will limit the potential business for the other growers.

If you see this about to happen in your part of the country, then it may be time for you to look at joining forces cooperatively with your competitors to start a pasture poultry producers cooperative.

Marketing & Business Management

Raising, processing, and selling pasture-based poultry is a good on-farm, family enterprise. Realistically, nobody is going to "get rich fast" doing this. While some growers have claimed $25,000 per year on 20 acres, the majority of growers are making a good bit less than that, and some are even losing money. This, like any other farm enterprise, requires good business planning, careful control of expenses, and a long term commitment to marketing.

Let's talk a little about sales. Many of us are just too shy to be good sales people. Don't let that stop you. After people get a taste of your chickens they will want more, and hopefully, a LOT more. Start with your friends and relatives and then branch out with sales activities to business associates, folks in your church, and so forth. There are many excellent books on how to improve your sales ability. Zig Ziglar is one of our favorite sales trainers. Check out some of his tapes or books and practice them. Being effective in sales will help you in other areas of your life as well. Salesmanship is worth studying.

What about samples as a sales aid? We rarely give a whole bird, as it is about the same a reaching in your pocket and giving away a $10 bill. A business card is a lot cheaper and will get the point across just as well. We donate chickens or turkeys to a local event if we know the product will be handled properly, cooked right and our farm gets credit for the donation. We also have done taste samples at farmers' markets or food fairs. This works very well, especially when serving the samples with something like home-made peanut sauce.

We find our average customer uses about 20 chickens per year. I think they would buy more, but most of them don't have freezers, except for the little one on top of their refrigerators. Because they don't have freezers they can't stock up. They will want you to store the chickens for them, and you will want to do that, but please remember to add that cost to your sales price especially in the off-season.

Also remember that most of your turkey sales will happen just before Thanksgiving. We started several years ago with 50 turkeys and grew to about 500 per year. Year after year we sent out a notice to all our customers asking them to sign up for a Thanksgiving turkey, but most of them wait until the last minute to call us.

Egg sales will continue year round, but you will need more in the spring and summer if you chose to sell at a local farmers' market. It seems that every spring there is an overabundance of eggs at the markets and then the volume drops as the summer wears on. Under proper storage, eggs can keep for several weeks so a large reach-in cooler might be a good storage addition to your operations.

Here are the approximate sales figures that will help you make some business decisions.

Broiler Sales: Many growers report that they market their broilers for between $1.50 to $2.35 per pound, dressed weight. Selling broilers for less than $2 per pound dressed will not give you sufficient profit to make this a good on-farm business for you. At the $2 per pound level, you can get a decent profit. Here's what you might expect to earn, using approximate figures:

Equipment cost	50-cents per bird
Chick cost	$1 per bird, includes postage
Feed cost	$1.90 per bird (16# at 12 cents/#)
Bag cost	10-cents per bird for broilers

Each broiler will cost $3.50, not counting your labor or overhead costs, and will sell for $8 (that's $2 per pound x 4 pounds dressed weight). The gross margin on each bird is then $4.50, from which you will need to earn a profit, pay salaries, land and building costs attributed to poultry and miscellaneous and marketing/distribution costs.

One person who is a steady and skilled worker will be able to process about 30 broilers in a good day from live bird to vacuum bagged and in the cooler. If you want to do 100 broilers per day then, you will need at least 4 hard-working people.

Egg Sales: Growers report sales prices from $1 dozen to $3 per dozen. Most of us are near the middle. At $2 per dozen, week in and week out, you can make a little money with eggs.

Here's a sample budget:

Chicks	$ 1.50 (includes postage)
Feed	$12.00 (100 # at 12-cents/#)
Housing	$ .50 per layer per year average
Egg cartons	$ 3.75 (25 cartons at 15-cents/carton)
Boxes	$ 1.00 per "30 dozen" box
Estimated total	$18.75

Income: 300 eggs per hen in a 60 week cycle = 25 dozen, selling for $2 per dozen = $50 per hen gross income.

Taking away costs of about $18.75 per hen, leaves you with a gross margin of $6.25 per hen. From that you will have to pay yourself a profit, salaries, land and building costs, and overhead.

Turkey Sales: Growers report prices of $1.90 to $3.25 per pound for turkeys at Thanksgiving. Most of us are near the middle. Here's a sample budget for turkey sales.

Turkey poults	$3.00 includes postage
Feed	$7.20 (60# at 12-cents/#)
Equipment	$ .50 per bird
Bags	$ .20 per bird
Ice	$1.00 per bird
Total cost	$11.90

Income: At $2.50 per pound, average dressed weight of 15 pounds, each turkey has a gross value of $37.50, less costs of $11.90, for a gross margin of $25.60. From that you will have to pay profits, salaries, land and building costs, marketing and distribution, and overhead.

Processing turkeys is hard. They are big and bulky, and it is really hard to get all the viscera out, especially the crop, windpipe, vessels and tissue along the neck. One hard working and skilled person can process about 15 turkeys per day from live bird to vacuum bagged and in the cooler. So, if you want to do a hundred turkeys per day the week before Thanksgiving, you will need at least 5 or 6 strong, hard working helpers.

To Insure or Not to Insure - Legal Ramifications

Insurance is basically risk management. Farming is constantly subject to uncertainty and risk. Natural disasters, fire, theft, diseases, and accidents can harm and even destroy your business and hard work. You will be doing yourself a favor by taking time to think about your business and its possible risks. You might ask: Can risk be eliminated or reduced by different contractual agreements?

By the way, one of the best ways to insure against law suits is to incorporate your business. We highly recommend you read: *Inc. and Grow Rich* by Allen, Hill & Kennedy. It's available at your local book store.

General Liability Insurance. We Americans have become the most litigious country in the world. People will sue over anything, so be prepared. At our farm we carry a $1,000,000 floater on our farm policy. It costs less than $100 extra per year. This covers any harm that might happen to clumsy visitors (and their even more accident prone kids), as well as our apprentices.

Product Liability Insurance. Many major stores and supermarkets require product liability insurance. We have talked with our insurance broker and she tells us it is available for probably $500 to $1000 per year depending on your volume.

Workers' Compensation. This insurance applies to all 50 states. It provides for employees to be covered for injuries suffered on the job. As an employer, you pay on behalf of your staff into a fund reserved for employee claims. In Virginia, any business with more than 3 employees must participate. This varies by state, so check with your state's office for your rates and requirements. Most small-scale farmers don't carry Workers' Compensation and failing to do so could be economically fatal if you, or one of your employees, were to be seriously injured.

You can sign up for Workers' Compensation through your local insurance carrier. You can also qualify yourself as self-insurers by submitting an application to the Workers' Compensation Division. Another alternative is to join or form a group of self insurers and file an application with the Department of Insurance for necessary approval.

There is a lot more about business management that we advise you to study, including: accounting, marketing, personnel management, tax laws, and cash management. Being a farmer is definitely a multiple-discipline profession.

We believe in continuing education as a part of our lives. Learning new things is fun and exciting for us. We like being on the steep side of the learning curve. Why not? We budget at least 10% for our continuing education, including books, workshops, tours, and anything that will expand our horizons.

One of our favorite slogans is: "If you think education is expensive, try ignorance." It pays to do your footwork before diving in head first into any business, project, or hobby.

POULTRY SELF IMPROVEMENT
Best Sellers
You Deserve More Than Chicken Feed
101 Great Grazing Techniques
Cross the Road and Never Look Back
Teach Yourself to FLY
Day Range Poultry
How to Stay Out of Chicken Soup - Forever!
Retire in Greener Pastures – on Your Side of the Fence!
Meetings with Remarkable Chickens
Scratch and Crow Rich
Crow with the BEST

Peck Your
Way to
the TOP

True
Grit

Chapter 4: Consider a Poultry Producers' Cooperative

Gene Logsdon, in a presentation he gave entitled the "Wild-eyed Wanderer", stated that "value–added" has become the darling of current agriculture trends, and this has a lot of merit. The problem with value-added is that the farmer must also plow, plant, harvest, process, and then market their crop. Anyone who has tried to do it knows this is the job of three different professions: a producer, a processor, and a marketer. You could literally kill yourself, and stress your family to the point of breakup trying to do it all.

We must face the fact that most of us are not supermen or superwomen. More often than not, value added requires more work from already overworked people and it often takes more than you and your family can give. Additionally, most farmers just don't have the personality to be a salesperson. We have to face the fact that any farmer, who is a real farmer, already has a full time job. If you then try to be a value-added producer, and after doing that switch hats to become a salesperson this is often the formula to stress and frustration. You, and your marriage, probably won't last.

Having said that, we believe that all farmers who are interested in growing, processing, and marketing poultry should be able to do so locally and profitably. They should be able to use locally produced organic feed, have their own processing facility, and have a customer base of loyal folks. How to do that?

This is where a producers' cooperative is needed: where the efforts of a group become greater than what any individual farmer might be able to achieve.

Competition Among Producers

The long term solution to unnecessary competition among Day Range growers might also be found in a producers' cooperative. This gives each producer two options: first, to sell directly to local residents. Second, to sell cooperatively to mainstream markets such as health food stores, supermarkets, and specialty stores.

Let's examine the local competition issue. Assume for that one grower can serve a population area of up to $25,000 people. It appears then that a dozen growers could operate in an area around a small with a population of 300,000. In reality, we think one or two growers in each area would thrive while the others would merely survive. The real probable outcome is that growers will compete intensely by lowering prices and offering more services to attract buyers. The end result is that none of the growers are able to sell for a good price and make any sort of living from it.

Processing Cooperative

The lack of processing facilities and necessary skills are major limits to growth for the pasture poultry industry. The cost to build a proper facility in order to get state or federal approval is quite high, too high for most small-scale growers. Processing skills are easily acquired with proper instruction, but speed comes only with practice and experience. In rural areas, it can be difficult to hire part-time processing help. Even in populated areas, it's difficult to hire capable people for infrequent need, and seasonal work.

Another area where a processing cooperative would be valuable is in addressing the legal quality of birds a Day Range producer can raise. For example, in about half of the states, there is a legal limit of 1,000 birds a farmer can produce without processing in a USDA approved facility. Above 1,000 birds, a USDA approved facility is required by law.

The 1,000 bird restriction has four major disadvantages for the individual grower:

1. The 1,000 bird production limit would only support a part-time job. The volume of sales just isn't there.

2. This low level of income prohibits the grower from buying and installing an adequate processing facility with necessary equipment to make the job more cost effective and labor efficient.

3. It might also severely restrict the grower to direct sales. In some states the "unapproved" product can only be sold to individuals who will cook it themselves.

4. The 1,000 bird production ceiling creates unnecessary competition between growers. If the growers can only sell to individual consumers, then two or more growers in the same locale will begin competing with each other for the same small market. This usually results in lower prices and yet lower profits.

With the collective buying power of a producers' cooperative, it will be possible to create a USDA approved processing facility with efficient processing equipment, including a walk-in cooler with freezer capacity sufficient to store large amounts of poultry. The cooperative would also acquire delivery equipment and would train staff. This facility could be staffed by the cooperative members or by paid staff who are able to work part-time and seasonally.

Marketing Cooperatively

Marketing can be a major obstacle for growers who do not have people skills and patience. The marketing aspects of the cooperative requires the most time, but also offers the greatest rewards.

In high population areas, there is a market for organic produce and there are plenty of enthusiastic and interested buyers. The difficulty is in accessing them. In rural areas, buyers are few and far between, and are often less conscious of superior food values in organically grown products.

Once a market has been established, it will be relatively easy to add products to the menu and to expand overall sales for every producer.

Purchasing Feed and Hatchery Cooperatively

With several growers needing chicks on a regular basis, another logical step might be to create a cooperatively owned hatchery. The hatchery will be able to produce superior chicks and poults with first generation hybrid vigor and genetic characteristics chosen specifically for pasture raising.

In most cases, individual growers have to finance their operations entirely by themselves. This is probably done without any access to financial assistance except through real estate, personal loans, or even — perish the thought — credit cards. It is true you can use off-farm income to provide start-up capital, but this may place an unrealistic strain on your household

The industry meaning of "free range": the door is left open at the far end of the barn.

budgets. Major funding from state, federal, private nonprofit or commercial sources to start the producers' cooperative is another possibility. With this assistance, the business could develop to a point where it provides a significant income to all of the growers who own shares in the cooperative.

Grower Guidelines and Standards

To help assure quality control, there needs to be clearly defined operating procedures on how the poultry is raised. This would include specifications about feed and rations, graze, and handling of both live birds and processed birds.

1. Poultry feed and rations. Ideally, this includes using locally grown grains. Long distance transport of grains from other regions uses tremendous resources and is not sustainable. The guidelines might also stipulate organic poultry feed.

2. Graze. We want to ensure that the poultry have access to green grass at all times after they leave the brooder. And the guidelines will also want to guarantee protection from weather, predators, and diseases.

The objective is to ensure that all growers use a superior method that will result in praise from their customers, a humane life for the poultry, and efficient management standards from the grower.

3. "Day Range" and What it Stands For. We are distressed that the term "free range" has become practically meaningless. It is much like term "natural". To comply with the current industry definition of "free range" only means the birds can wander around in an over-crowed barn.

Our intent and part of the momentum behind this book is to have the words "Day Range Poultry" actually signify a quality, premium, custom-raised product that merits the high price customers will pay for clean, humanely raised meat.

Our intent is that any grower who raises poultry by the Day Range method could have the advantage of using the logo and wording to help sell their products throughout the US and even internationally. Charles Ritch suggested we use the drawing from the cover as a logo to identify this way of raising poultry.

The new national logo for "Day Range Poultry™"?

This is still developing, so stay tuned to the web poultry discussion groups — especially the dayrangepoultry list on yahoogroups.com.

This brand identification ("Branding") is done in France, for example with the Label Rouge (red label) poultry. These slower growing, forest and pasture birds are raised in non-intensive, more natural setting. They are also darker colored and usually red feathered birds.

The Label Rouge symbolizes wholesome excellence, gourmet flavor, humanely raised and processed premium poultry. Currently there are about 8 million birds raised this way in France, many of which are exported to England. The taste and texture of these birds is truly distinguishable, and they often bring as much as 100% higher price.

Chapter 5: Designing Day Range Shelters

As you can see from the many photos in this book, there are a lot of ways to design and build shelter structures to use in your Day Range system. The last chapter described some of our mistakes and lessons we learned. We believe the ideal building for day ranging has yet to be created. Its structure will vary depending on climate and terrain. What we explore in this chapter are the things you need to consider before you build. Throughout the section we use the terms turkey/chicken tractor, pen, shelter and skid interchangeably (or, to describe the shelter part of the Day Range system.)

Size. First of all, make sure your shelter is large enough to handle the size of the flock you intend to grow. The shelter needs be large enough to grow the poultry full term just in case you can't let them out on pasture. In "confinement barns" growers allow about 1/3 square foot per broiler. In day ranging, we allow about:

- Broilers – 1.2 square feet of shelter space (with at least 5-square feet of pasture)
- Laying Hens – 1 square foot per hen
- Turkeys – 4 to 5 square feet per bird

This way, even if the weather is totally inclement, we can safely grow the birds to maturity. Because they still have sunshine, fresh air, and exercise, they will taste far better than supermarket birds, even if they never touch grass.

The foremost purpose of the shelter is to protect poultry from sun, rain, and predators. It needs a good top that won't heat up to the point that your birds get overheated. The shelter needs sides and ends that will keep out predators, rain and drafts, while letting in fresh air and sunshine. The shelter needs a floor or deep bedding piled on the ground so the birds are protected from cold ground, dampness, and rainy weather.

We prefer to use inexpensive materials that are lightweight and easy to put together, yet are heavy enough to anchor the pen in strong winds. We make our shelters easy to move, either by hand, with a tractor, car, pickup truck or ATV.

Our most successful shelters are on skids. These skids are made with pressure treated runners (skids), galvanized steel greenhouse bows, 22-mil woven poly cover, and chicken wire on the ends. The floor is pressure treated plywood, covered with a 4-inch layer of planer shavings that we get free from our local Virginia Horse Center. The horse manure in the bedding does not harm the poultry and the bedding is free for the hauling.

Design Factors for Turkey Day Range Shelters

The ideal turkey tractor skid is very similar to that for chickens, but the turkeys prefer to roost as high as they can. Turkey skids need the following design characteristics:

- Critter proof
- Water resistant
- Moveable
- A roof high enough you can get inside as needed
- Ease of getting the turkeys in and out
- Cost effective
- Has above ground roosting space and
- Room for water fountain and feeder if needed

In the case of breeder flocks, the ideal skid will also need one or more nesting boxes and separate paddocks for the toms during mating season if you are pasture breeding.

If you allow a minimum of 5 square feet per turkey, then a 128 square foot skid (8-feet x 16-feet) will hold 25 full size turkeys. This assumes that you will be harvesting the turkeys before they get much larger than 20 pounds each. In a pinch—say for overnight protection—you could crowed as many as 60 turkeys into this same skid. They would be very crowed and likely to peck at one another and fight if they are held this way more than a few hours, or overnight while they are mostly sleeping.

In some rotations you may be able to use the same skids for two flights or flocks of turkeys during a growing season. You could grow the first batch during July and August for a total of 8 weeks on pasture. Then you harvest that flock and either sell them or put them in the freezer for Thanksgiving sales. Keep in mind there is only a limited market for turkeys at other than holiday seasons.

You will need a walk-in freezer to handle this many birds, since chest freezers just can't hold enough turkeys to make it economically feasible, unless of course, you already own the freezer. Then you put the second batch in the skids beginning in September and harvest them for Thanksgiving.

Construction Materials

We first ask: what materials do we already have and are paid for? Since the skid will be in contact with the earth and will be exposed to the elements year-round, then we want to use materials that will last the longest. In many areas locust or white oak boards direct from the sawmill will be the best choice. These native hardwoods withstand weathering and ground contact. However, they are heavy and many sawmills won't have locust since the wood is so hard to work with.

You may wind up using pressure treated southern yellow pine. Many growers stay away from pressure treated boards because of their supposed detrimental impact on the environment, both in manufacture and in use. However, when you compare the longevity of pressure treated boards against having to rebuild the pens at frequent intervals, the pressure treated lumber may be the least costly to the environment and to your pocketbook.

Construction lumber comes in 2-foot increments, with plywood coming in 4 x 8-foot sheets. Since a 10-foot wide structure will fit through our barn door, and 16 feet is a good length for boards, we decided to build on skids 10 feet wide and 16 feet long. This gives us 160 square feet, in which we can grow up to 80 turkey broilers for up to about 20 weeks, before we start harvesting the larger ones for the freezer.

How many birds (young or mature) will you be holding in each skid? If you are using the skid as a temporary brooder house, then allow 1/2 square foot per turkey poult, up to about 3 weeks of age. Allow 1 square foot for growing turkeys to 10 weeks, then 2 square feet up to 24 weeks. These square footage calculations are based on nighttime confinement only, with the birds having unlimited access to the outdoors during at least 14 hours per day.

If you see signs of fighting, cannibalism, or feather picking, try giving them more room. Some fighting between toms is normal, but injury often occurs especially during breeding season when toms are combatting each other for the favors of the ladies. It is necessary to separate toms during the breeding season. You can get a better hatch rate by switching toms amongst the hens each week or ten days. We only let 1 tom in with about 8 to 20 hens at any one time. If you let 2 toms in, they will fight and knock each other off the hen when mating, and the fertility rate drops.

Is Your Land Rocky, Hilly, Rough, or Level?

Will you be using the skid in the summer and winter? Will you be using it for a brooder space as well as for a growing out space? Will you be moving the skids by hand, or do you have a farm tractor or other vehicle to move the pens?

If you build your turkey tractors inside a building, are there any limits on door openings and so forth that would hinder your ability to move it out after it is built? We've all heard the story about the guy who built the row boat in his basement and then had to cut it in half so it would fit through the door. I've never done anything quite that bad, but I did once build a wall in sections on the floor, only to find that I couldn't stand the sections up and had to take them apart and rebuild them in place.

There really isn't just one way to design or build a turkey tractor. It depends instead on what your specific needs are. For example, if you have to move your portable pens by hand, then you will necessarily want to build them small, light, and perhaps with wheels. This may mean the pen will be too light and too small to handle more than just a few birds, especially at maturity. Conversely, if you have a farm tractor available, then your pen design can be much larger and more heavy-duty.

We build our turkey tractors inside a pole barn at our farm. The barn has a free span of thirty feet, so we can build any length up to about twenty-four feet and still have room to work around all sides. However, the sliding door is only eleven feet wide and ten feet high, so this limits how wide and how tall we can build a structure and still be able to move it out of the barn. We have a thirty-five horsepower, four-wheel drive tractor to move the skids, so the weight or size of the turkey skid is not so important.

Because our land is mostly sloping with some terraces, sink holes, and quite a few rough spots to cross, it is necessary for us to build the pens more ruggedly than they would need to be if the only criteria was just to house turkeys. We move the pens from one spot to the next, and sometimes over uneven ground and rock outcroppings. This necessitates a rugged, well braced, strong structure.

We also have opossums, raccoons, foxes, and large neighborhood dogs to contend with, so we put chicken wire or roofing panels at any place predators might try to gain access.

If the skid is totally enclosed, you might want to include a pop-hole in each wall for releasing birds into certain parts of the paddock. Or, in the case of chickens, let them out during the day and close them in at night.

Chapter 6: Egg-ribusiness

This chapter discusses the management of eggs to sell for food. It includes sorting, grading and storing, layer hen management for optimum egg production, and how to be "egg-stra safe" with handling, as well as cooking tips. We discuss incubating and hatching in the Chapter entitled: On Farm Hatchery.

Hens begin laying eggs at five to six months old. Initially, eggs tend to be smaller, oddly shaped, and with various patterns on the shell. These are beginner (pullet) eggs. As the hen matures, her eggs become more normal in appearance.

An egg contains all the nutrients needed by a developing bird which explains why it is so nutritious. In the egg production system, an egg starts as a small ovum attached to a yolk in the hen's ovary. The yellow yolk is derived from precursors synthesized by the liver. The passage of the egg and yolk stimulate the wall of the oviduct to secrete successive layers of albumin or egg white.

When the eggs reach the lower contracted part of the oviduct, the last two tenacious thin layers of albumin are added to form the inner and outer shell membranes. Just before the eggs are discharged into the cloaca, the shell gland adds the shell. This shell is initially a coating of thin liquid secretion that contains lime, which hardens the shell quickly.

The egg is laid about 24 to 36 hours after leaving the ovary. It is damp when it is first laid and dries quickly. There is a protective water soluble coating on the egg that helps keep the egg clean and bacteria out. The shell is extremely porous and air permeable, in order to supply oxygen to the embryo during incubation.

We collect our eggs once or twice daily. If an egg is left in the nest too long, it gets trampled, dirty, and possibly broken. Broken eggs are especially bad. Besides losing an egg (and profit), this also makes the other eggs dirty and harder to clean. It can also be a temptation

Seeing a nest of eggs brings out the mother even in Atilla the Hen. Once she goes broody and sits on a nest her egg production drops.

for hens to begin egg eating. The longer an egg is in the nest, the more likely it is to get soiled with droppings.

Many eggs in a nest can bring out the mother in any hen. The sight of a clutch of eggs can put a hen in a broody mood and she will set on the nest. This decreases her egg production.

In warmer weather a fertilized egg begins developing quickly. Within 3 days at the right temperature (99 degree F.), an embryo is developed enough to be bloody when the egg is cracked open. No one wants to crack open an egg that is partly developed — YUCK! Selling eggs with partly developed embryos is a good way to lose customers and get an undesirable reputation fast.

We use egg baskets to collect eggs. We only fill the baskets half full with eggs or else the bottom ones tend to break from the weight. It is important not let the eggs knock against each other and get damaged in transport.

Take care of any cracked (even hairline cracks) or soiled eggs immediately; they will turn rotten fast. We use some of our damaged chicken eggs to feed back to turkey pullets. This increases the protein in their diet. We freeze the eggs, cracked shells and all, in one quart wide-mouth containers (somewhat like large yogurt containers). Then we just put the frozen eggs (still in the container) in with the young turkeys and they feast on them.

No, feeding chicken eggs to turkeys is not cannibalism. The eggs are from a different species (chicken) and is some of the highest quality protein Nature makes. However, we don't give shells to our future breeder pullets because we don't want them to learn about eating egg shells. We buy our breeder flocks for producing eggs and we need those turkey eggs for hatching.

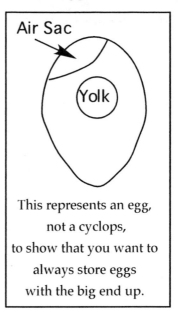

This represents an egg, not a cyclops, to show that you want to always store eggs with the big end up.

Grading, Sorting, and Storing Eggs

We sort our eggs and get them into refrigerated cool storage as soon as possible. Store eggs with the large end of the shell up. This helps to keep the yolks centered in their shell and the air sac to form at the top. It also makes the eggs look bigger in the cartons.

Store eggs away from strong odors. The shells are porous and will absorb any strong smells, especially onions and garlic. We mark the egg cartons with a code for "lay date" so we will know which ones are the oldest and to sell them first. This is the classic FIFO, or first-in-first-out, inventory control system.

Grading Eggs. All of our direct sale eggs are sold and labeled as unsorted and ungraded. This allows us to mix the smaller ones in with the jumbos. Chefs and some of the more upscale food stores want only large, extra large or jumbo eggs. To sort these we use a simple egg scale and deliver these eggs in flats with 30 dozen to a case.

Market Sizes for Eggs (approximate weight of 1 dozen)

Jumbo	30 ounces
Extra Large	27 ounces
Large	24 ounces
Medium	21 ounces
Small	18 ounces
Peewee	15 ounces

Adopt a Hen - Increase Egg Sales

Every hen needs a home. We had a lot of success with this promotional idea which we picked up from the Internet. Of course, you keep the hen on your farm and the customer gets eggs from any of your chickens you have employed.

We didn't really keep track of each hen's parents, but if the kids wanted to know which one was theirs, we asked what they named her. We then could call her by name, and catch any hen that is close by and let them hold her. They are thrilled.

What this program does is sell pre-paid eggs. This gives you cash upfront and guaranteed eggs customers. Your customers get lots of the BEST heart-healthy eggs at a great price. The customer gets to name their pet chicken after their favorite Greek Goddess, movie star, or dearest relative. This becomes a fun entertaining conversation item for them.

We give our customers a certificate of adoption, suitable for framing. We use the certificate above and change the chicken's and owner's names. The actual size of the certifi-

Adoption Certificate

This certifies official adoption of the hen named

Marie Elizabeth

Identification Number 27

Adopted by
Scooter and Joe McMoneagle

With all associated egg rights guaranteed

This hen free ranges on organic pasture, and
eats only the finest wholesome feed & bugs avilable.
She resides in the layaway at

GreenWay
20 GreenWay Place
Buena Vista, VA 24416
(540) 261-8874

Thank You For Supporting Local Sustainable Agriculture

cate is regular paper, 8.5 by 11. It only takes a few minutes to print the certificates from our computer. As an added incentive, we also give them five gallons of the finest organic composted chicken manure for their gardens.

Prices vary, but this is an idea about how we bundled our Adopt-a-Hen egg prices:

- Plan A: 40 dozen eggs and 10 gallons of compost for $80.00 ($2.00/dozen)

- Plan B: 20 dozen eggs and 5 gallons of compost for $43.00 ($2.15/dozen)

- Plan C: 12 dozen eggs and 1 gallon of compost for $27.00 ($2.25/dozen)

Hens Eating Eggs and How to Stop Them

Hens destroying eggs are literally destroying your profits. We figure each egg is worth at least 18 cents. So if you are losing even one egg a day this totals about $65 loss per year. Egg destroying by hens is a very serious problem that can run through your flock like wildfire. One of our friends had an egg eating hen. Soon all the birds began doing it. He had to put them all down — from egg layers to BBQ.

There are several steps you can take to prevent or stop it. Here are ten preventative actions you can routinely do:

1. Always have enough oyster shell available. Sometimes hens will eat the shells as a source of calcium. Poultry eggs take about two weeks for shells to begin to thin from the lack of calcium, converse-

Make sure your hens have enough to eat. Bored, hungry hens are more likely to peck and eat eggs.

ly, once the hens have access to calcium it takes about two weeks for the shells to thicken.

2. Make sure your hens have enough to eat. A hungry hen looks for food, and eggs are quite tasty.

3. Dim lights or slatted curtains across the nest box to make the nest darker.

4. Large roomy nests – so eggs don't pile up and get smashed as easily. Lock the hens out of the nest at night so they don't roost in there and crush the eggs.

5. Never feed shelled eggs back to the chickens. This will teach them to be egg eaters.

6. Remove cracked eggs from the nests and don't throw the shell down for the hens to eat. Once a hen learns to eat eggs the habit spreads through the flock quickly.

7. Collect eggs frequently, at least once daily or more often.

8. Keep your birds from getting bored by giving them a variety of garden scraps, range, and space can help.

9. Roll away nest boxes get the eggs out of the hen's beak-reach almost as soon as they are laid.

10. Remove any broken shell fragments, or egg contents as soon as you see it. Never let shell fragments accumulate, as hens will quickly learn they are a source of calcium.

Remedies for Egg Pecking – Don't Delay

Once you have egg pecking for any reason, be sure to take quick action to find the cause and correct it. If you do nothing, it will get worse and cost you money. Luckily, we have always been able to stop our hens from egg eating. Here are a few solutions we have successfully tried:

- *Close the Nest.* Once we find a nest box with a pecked egg or cracked shells in it we remove that nest box from hen access. To do this, we either take the bottom out, or block the entrance.

- *Busted!* If we catch a hen eating an egg we remove her from the flock for a few days. This tends to break the hen's normal routine. In poultry psychology terms, this is using Poultry Neuro-Linguistic Programming, or PNLP, to change the bird's state and focus.

We also band her so when we put her back with the flock, we know who she is. If we catch her eating eggs a second time, then we cull her from the flock permanently.

- *Rats and Possums Love Eggs.* If you are finding egg shells in the early morning, then you just might have fed the wildlife over night. If we suspect a rodent problem, we will visit the barn or layaway after dark. If we see any rodents, we begin setting traps. In the early spring when we keep our layers in the barn, we once found a possum in the nest boxes munching on an egg laid after we had done the daily collection. These possums had somehow gotten in and enjoyed sushi omelets for over a week before we figured out what was going on.

- *Flock Swap.* Finally, if the egg destroying situation is so bad that you are losing all your profit or even worse, losing money, then it might be best to start over with a new flock and try not to let it happen again.

Common Egg Problems You Might See

Double Yolk. This is common with large eggs and is the result of two yolks being released into the oviduct at the same time and enclosed by the same shell. Sometimes double yolks are a sign that your hen is too fat. When the oviduct contracts there is not free passage of the ova and two yolks pass at once. While our customers tend to be thrilled with double yoked eggs, the eggs are often so large that they get cracked in the cartons.

Egg with blood on the shell. Young pullets just beginning to lay eggs sometimes have to strain and cause tears in their vents. An overweight hen will have a hard fat-deposit just below her vent. If a large egg is formed it will actually break when the hen is trying to pass the eggs. Sometimes such hens will end up with a prolapsed uterus and suffer a prolonged death. You can tell if the egg is breaking inside the hen if her rear feathers are covered with egg white and yolk. This can also lead to vent picking where other hens will pick at the rear of the unfortunate hen. See *Vent Picking* for more information.

Green Yolks. Dr. Seuss was not kidding with green eggs and ham. Sometimes a range hen will eat acorns or shepherds purse. Both of these can cause green egg yolks.

Pale Yolk. Rich, yellow yolks are the result of the hens getting green forage. In winter, when the grass is not growing and hens don't have as much access to greens, thus can result in pale yolks. Feeds with grass meal and corn help improve the color. Frank Perdue says he feeds marigold petals, as a health food, to help give the yolks that yellow glow.

Malformed & Mid-ribbed Eggs. Some eggs might have uneven, ribbed, wavy shells, soft ends, and odd shapes. These are usually from younger hens. It can also be caused by a diseased condition of the lower part of the oviduct where the egg is forced through a small opening and, because the shell is soft, it is distorted into a peculiar or elongated shape. It might also be due to a disease such as bronchitis or the viral caused egg-drop syndrome. It can also be due to a lack of calcium, so make sure the hens have enough in their diet.

No Yolk. This is often found in a pullet's first few eggs, or as a sudden shock to an older hen. The problem is usually temporary.

Nude Eggs. This is an egg laid without a shell. This problem is usually temporary and happens with pullets, or hens that are stressed, or have a sudden change in temperature. Make sure oyster shell is readily available.

Extremely Small Eggs. These are common at the beginning or end of a laying period. This is due, in part, to a smaller size and decreased secreting power of the oviduct.

Egg-bound Hen. This is the inability of a hen to pass the egg normally, that is, egg constipation. The affected hen becomes restless and makes frequent visits to the nest in an effort to lay. The straining can cause inflammation and sometimes a prolapsed (fallen down) oviduct through the vent. Other hens can be attracted to the protruded part, peck and tear out portions of the hen and causing her great suffering and a painful death. Once we see the oviduct has prolapsed we simply humanely put down such hens, as we have yet to get one to return to normal.

Another symptom of egg bound is a hen with a dirty rear end. This results from the egg breaking in the oviduct and causing a messy hen. Instead of egg on their face they get egg on their butt. You can imagine how hard that would be to deal with without toilet paper!

Egg bound can easily happen in the broiler breeds you keep for breeding stock. These birds are genetically bred to gain weight fast. Once mature, they continue to fatten up to mega Rubinesque proportions if you don't restrict their feed intake. On our first broiler-breeder flock, we did not restrict the feed enough and almost 100% of the hens became egg bound because they were too fat.

In young pullets, the usual cause of egg bound is the attempted passage of a double yolk egg before the egg passage has become sufficiently dilated. If you really like this hen and want to treat her, then the surest treatment is to remove the egg by lubricating the oviduct with mineral oil or petroleum jelly. Then passing your forefinger through the vent to guide the egg while at the same time press with your other hand on the abdomen. This forces the egg through toward the vent.

Once you see the shell in the vent, puncture it with something sharp (be careful here) and withdraw the egg parts. Putting the hen in a warm place will sometimes let the egg come out on its own. The warmth helps the muscles in the oviduct to relax.

If the cloaca is inflamed, you can inject cold water to ease her discomfort. If the oviduct has been inverted through the vent, clean and push the parts back and inject cold water frequently. Sometimes, in younger hens, the oviduct will remain in place. A hen that gets plenty of exercise and good nutrition rarely develops this problem. The likelihood of permanently curing an inverted oviduct is remote. In most cases we quickly and humanely dispatch the affected hen and concentrate on the healthy birds in the flock.

Shell Color. Egg shell color varies from pure white to many shades of browns, deep olive, and even turquoise. Araucanas, a breed named after an Indian tribe in Chile, is called the "Easter Egg Chicken" because of its naturally colored egg shells. Shell color is due to a breed's genetics and not due to feed. Brown and colored eggs have a surface pigment. If you rub a newly laid wet egg the pigment will rub off. Once the egg has dried the pigment is set. We haven't seen any reports that claim one color is more nutritious than another, but in our area brown eggs outsell white eggs at least two to one.

Yoke Color. Yolk is the Old English word for "yellow". The color of the yolk can be greatly modified by the hen's diet. The inclusion of oxycarotenoids will produce a deep orange color. Hens with access to grass tend to have deep yellow or orange yolks.

Fertile vs Infertile Eggs for People Food

In certain markets, fertile eggs bring a higher price. Some folks think fertile eggs have more vitality or strength. However fertile eggs can be extremely costly to the producer. As soon as an egg is laid the embryo begins developing and it develops fast. This can especially be a problem in the summer if you for some reason wait a day or two to sort and refrigerate the eggs. That day or two delay could result in embryo development, and your customer getting a rude surprise when they crack open the egg to find a bloody mass when they were expecting a beautiful organic egg.

A hen produces the same number of eggs regardless of a rooster's presence. And if hens had the right to vote, they would probably vote not to have that cocky guy around causing so much chaos. Larger, dual purpose roosters especially can be vicious to the hens and cause injuries. Producing infertile eggs (no roosters with the hens) for human consumption eliminates the danger of all embryo development.

The advantages of producing infertile eggs are:

- They do not hatch.

- They will not develop into a chick (and surprise your customers).

- No little unappetizing blood spot on the yolks.

- They ship well and are more easily preserved.

- They are slower to decay.

- Roosters are not required – less feed is consumed.

A hen without a rooster is like a fish without a bicycle.

Gloria Steinahen on the usefulness of roosters.

We find that visitors to our farm love to see and hear roosters. It is part of the farm experience. The males are flashy and provide some farm-fun photo opportunities. So in our 350 hen layer flock we keep one or two roosters, usually small bantams. With this ratio of rooster:hen we know those guys can't get around that much and the Napoleon-complexed, bantam-boys are often chased away by the larger, uninterested hens.

Candling Eggs for Quality Control

Sometimes we find eggs in odd places and we are not sure about how old they are. In this case we look at the egg through an egg candler (egg tester) that shines a strong light through the egg. This lets us examine the interior. Candling eggs is simple. Put the large end of the egg against the light hole of the egg candler with the small end extending downward at about a 45 degree angle. With a twist of your wrist give the egg a quick twirl and observe the egg contents. We cull any egg that has blood spots, embryo development, dark colored goo, or watery whites. A large air sac indicates the egg is older and we don't sell those to our customers. Usually, the eggs we find in unexpected places wind up as pig or pet food.

We handle eggs to be incubated differently and very carefully. We never twirl the egg to examine the inside. A quick sudden movement like that will damage the sensitive membranes and cause the chick not to develop. See the chapter on incubation for more information on candling and how to make your own simple, inexpensive candler.

Cooking with Eggs – Be an Egg-spert

It is hard to find a handier food than an egg with its unit-of-use natural packaging. Of all foods, eggs are among the most adaptable and nutritious. Adored by chefs, with a shape artists and engineers admire, and an economy that frugal folks crave, eggs are a universal wonder food.

In cooking, eggs serve a wide variety of tasks from the elegant presentations of the souffle, coating bread crumbs for frying and holding meat loaves together, keeping oil and vinegar from separating in mayonnaise, and crystals from forming in candies. They magically spin into meringues and are without equal for thickening smooth custards. They transform cake batters by providing a structural framework and produce finely grained ice creams, clarify or enrich soups, and glaze rolls and pie crusts.

There is a saying among culinary schools that "you can tell great chefs by the way they cook eggs". No matter how much in a hurry you are, don't rush cooking an egg. Cook them at a low-to-moderate temperature. High heat sets the protein in the yolk and white too rapidly and causes it to shrink quickly. No matter if you scramble, poach, bake, hard-boil, or fry the egg, cooking it fast will cause it to become tough and rubbery. In a sauce, an egg cooked too quickly will not hold the liquid properly and your sauce will curdle with small, tough lumps of egg. If the egg is in a soufflé, meringue, or angel food cake, the protein will not expand and your finish masterpiece will lack volume and lightness.

You can easily test the freshness of an egg by putting it in a bowl of water. Eggs that sink are fresh because the air sack has not yet formed inside the shell. If an egg floats it is older. Rotten eggs swim at the top. Be careful. Such eggs can explode!

A fresh, pasture-raised egg has a deep yellow yolk that domes up and stays up. The egg white is thick and translucent. A small dark spot on the egg yolk indicates the egg was fertilized. Remove the fleck only if you are making a light colored sauce or confection. Another test chefs do is to fondle the yolk. A fresh yolk from a healthy hen will not break when you hold it up and even pinch it by two fingers.

Hard Boiled Eggs. Put eggs into cold water first and heat them until the water boils for about 3 minutes. If you use eggs that are three days or older, the shells will be easier to remove. The shell peel-ability is affected by the egg's pH. A fresh egg has a pH of about 8. A three day old egg has a pH of about 9.2. That's why some cooks say to put salt in the water to boil eggs because it raises the pH to more basic. Another tried and true way to get your hard boiled eggs to peel easily is to put them in cold water immediately after taking them off the heat.

We prefer to steam our eggs, which also makes the shells easy to remove and seems to make a tenderer "hard boiled" egg.

Not sure if an egg is hard-boiled? Just twirl the egg. Hard cooked eggs will spin like a top. Uncooked eggs will wobble like Humpty Dumpty, and might start spinning again if they are stopped abruptly. This is because the liquid inside a raw egg has an inertia of its own.

Occasionally you will get a hard boiled yolk that is greenish-gray discolored. This color is due to the formation of ferrous sulfide (FeS), a harmless compound of iron and sulfur, that is formed when the egg is heated, especially with older eggs. The yolk contains a lot of iron and the white has sulfur. The way to minimize this discoloration is to cook the egg only long enough to set the yolk and then put the cooked eggs immediately into cold water and peel the eggs promptly.

Exploding eggs are due to a build up of hydrogen sulfide (H_2S). In small amounts hydrogen sulfide gives the pleasant odor to cooked eggs and meat. In large amounts, you get the smell of rotten eggs that is so offensive.

Be Egg-stra Safe!

For many reasons, salmonella poisoning is becoming more of a concern. Salmonella is naturally present in the GI tract of healthy humans and chickens. However, one problem is that it can be passed from a healthy chicken directly into an egg no matter how clean the environment. To avoid food poison bacteria try to:

- Avoid eating foods that contain raw eggs, including cookie dough and batters.

- Fully cook foods that contain eggs. Salmonella is killed at about 160 degrees. If you prepare fried or poached eggs, the yolks will be firm when fully cooked.

- Store raw eggs in their original carton in a colder section of the refrigerator. Do not freeze.

- Generally, keep hard boiled eggs only about a week.

Egg Trivia

A hen requires 24 to 26 hours to produce an egg. About thirty minutes after she lays an egg, the cycle begins again.

An egg shell has about 17,000 tiny pores on its surface. That's why the egg can absorb flavors, odors and bacteria.

White shelled eggs are produced by hens with white feathers and white ear lobes. Brown shelled eggs are laid by hens with colored feathers and colored ear lobes.

Egg yolks are one of the few foods that naturally contain Vitamin D.

Yolk color depends on the hen's diet. Day Range hens lay eggs with rich golden colored yolks that come from the greens and chlorophyll in their diet.

This shows just how much our customers like the eggs from our organic, Day Range hens.

Chapter 7: Day Range: Fencing the Flock Not the Field

Day ranging poultry is a hybrid combination of mini-confinement and the free ranging system. In the confinement model, the poultry are held inside a barn for their entire lives, never having access to fresh air, sunshine, or green grass. The birds are protected in this way from predators and weather extremes. The disadvantages are that the barns are expensive to build and maintain and the poultry often require high maintenance including veterinary support.

On the other hand, free ranging poultry, at least in the old style, simply meant turning the poultry out into a field, with a range shelter they could go to during bad weather. However, to operate a free range flock in this day and age may not be possible because of the number of predators that we have to deal with, and the unacceptably high number of mortalities caused by weather extremes.

This outdoor model should not be confused with large, commercial poultry enterprises that can claim "free range" because of faulty USDA definitions. In this case, the birds are free to roam about – but are still confined in an artificial environment and overcrowded situation.

What we need is a poultry production system that protects the birds from weather and predators while giving them access to sunshine, fresh air, green grass, and plenty of exercise, and at the same time regenerates topsoil. And that's just what the Day Range system does.

We use inexpensive yet sturdy shelters with floors to protect the poultry from predators and inclement weather, and we use portable electric poultry netting to contain them inside a temporary paddock. The fencing is the brake to keep the poultry in and the predators out.

Fencing Materials

The key to successful pasture based poultry is portable electric poultry netting. It is inexpensive, quick and easy to move, predator proof, flexible on uneven ground and rough terrain, and makes multi-cropping and multi-species grazing nearly effortless.

This kind of poultry fencing has been used for several years in other parts of the world, but has only recently become popular here in the United States. It is now available from at least two major suppliers in the US: Kencove and Premier (see resource guide). Both companies have excellent catalogs that will guide you through the steps required to build a good enclosure for your poultry.

Electric poultry netting comes in several configurations, ranging in heights from 36-inches to 48-inches, and lengths of 85 feet or 165 feet. The main difference between various models of poultry fencing has to do with post spacing, stay, and horizontal spacing. The posts are usually plastic with spacing from 10 to 12 feet apart. They are threaded into the netting and have either one or two steel spikes for stepping into the ground. We first learned about the British Electronet from our Australian friend, Michael Plane. Kencove is the American distributor for Electronet. We prefer the British made double spike "step in" model, which is much easier to install in the ground and holds the fence more rigid.

The "stays" are the vertical strands that help hold the fence upright. They are generally not electrified, and can either be woven or braided or flexible plastic. The horizontals are the parts of the fence that are electrified. True poultry netting will have lower horizontals at 3.5-inch spacing and stays at 3.5-inch spacing. For larger birds such as turkeys, large broilers and layers, you can also use a lighter, less expensive netting (NP-7) from Kencove that has stay spacing at 7-inch intervals.

Prices per 165-foot roll range from $130 to $200, depending on model and supplier. This may seem expensive at first glance, but you can fence an entire field, one day at a time, with just one roll of net.

Each poultry net has metal tabs at the end of the roll for connecting to the charger and to each other. You can use the nets individually to enclose an area of about 1400 square feet, or you can connect two or more together for enclosing larger areas.

We use the netting for goats, beef, broilers, layers, and turkeys. However, if your intent is to just use netting for livestock, then you can purchase less expensive rolls of netting that are typically called sheep fence. They are only 27-inches to 36-inches high and have larger horizontal and stay spacing. These sheep nets are also a lot easier to install and move than poultry netting, because the larger spacing helps make the nets very lightweight and easy to use.

You will need an electric fence charger. These come in a wide variety of models running from 120-volt plug-in electric, to solar and battery powered systems. Because of the portability of our poultry pens, we use a battery powered charger with a deep cycle marine battery. The battery will hold a charge for several weeks, then we recharge it with an ordinary battery charger.

We use a lead wire running out from the charger to connect to several fences. At any given time we will have six to ten flocks of varying ages, each having its own poultry netting. This way we can keep the flocks separated and move individual flocks as needed for pasture management. One battery and charger will run five or six poultry nets. We always mow ahead of the pens so that grass and weeds don't touch and ground-out the netting.

Kencove also has a British made "pos-neg" or "positive-negative" model that is made for use in dry terrain where ground rods are not effective. The net spacing is 3.5-inches, and the roll is a few pounds heavier than ordinary poultry netting. It costs about $35 more per roll than ordinary netting.

Installing poultry netting is fairly easy if you are careful to follow the instructions in the catalog and in the shipping container. Here are couple of tricks we've learned that makes the job easier.

Tips and Tricks to Installing Poultry Netting

As with most things, the directions don't tell you everything. Here are a few lessons we have learned while working with several models of poultry netting.

First, mow ahead of the fencing either with livestock (grazers) or mechanical means. This helps decrease "grounding out". It also cuts weeds or shrubs that can tangle in the fence making it harder to install and take up.

Second, lay the fence in a circular pattern on the ground beginning at your power source and ending in a loop, or at the point where you will tie in your next roll of netting. To do this, secure

the first post in the ground and then walk around your pattern, allowing each next post to be pulled out of your hands as you proceed. If you collect the fence as we describe below this will be an easy task. If you don't pick up the fence in a systematic way then this can be an entangled mess. It is worth taking the time to do it right in the first place.

Now, go back to the first post and step or push it in the ground. Go on to the second post, pull the fence tautly from the first post and secure the post in the ground. To help drive the spike in, we found that by placing your knee in line with the fence and using it as a pivot point to stretch the fence as you step the second post into the ground. Then proceed to the next post, and so on.

By stretching the fence as tightly as you can at the bottom and letting the top of each post lean out about 10-degrees, you will get a tighter fence that does not sag so much. If you find that the posts want to lean in as you go around your circle, you may want to install temporary step-in posts along the fence to help tighten it. Another way is to use tent pegs and a temporary tie string to brace the fence, especially at the corners.

In hard ground, such as during a dry August, it will be difficult to get the spikes into the ground. Put your weight on the step-in, then wiggle the post from side to side. This will usually work. If you are using single spike posts you'll have to bend over, brace your knee against the post, and push it in with your hand while wiggling it back and forth — sort of prying it into the ground.

If the ground is too hard though, you may find it necessary to use a steel bar and hammer to make holes for the fence posts to sit in. We have to do this when we set fences near our driveway and parking lot where the gravel is packed hard from traffic. If you have a need to set a fence on hard ground frequently, as when you are grazing your road shoulders, then install one-half inch pipes at proper intervals for the step-in posts to rest inside.

I would often see Patricia going to move fences with her hammer in hand. This was helpful to drive the double spiked posts in by hitting the bottom. Don't use a hammer on top of the plastic posts without covering the end with a protector or the post will break.

Electric poultry netting will only work if you have a good charge in it. From the very first day you want the poultry to get SHOCKED the first time they touch it. Once they learn what it means, they'll respect it and stay clear of it. However, if you put them out there without the fence charged, they will simply come and go at will. Then when you do turn the fence on they won't respect it.

Broilers are easy to hold inside electric poultry netting. They move fairly slowly and don't fly. You can keep layers and turkeys from flying over the fence by clipping the primary wing feathers from one wing. The clipping needs to happen at an early age before they learn they can fly. By the time the primary wing feathers have grown back the layers and turkeys will be trained to the fence and will generally stay put.

You might also have a few chickens or turkeys that still get out. They will hang around the pen looking forlorn and it's pretty easy to get them back in at night.

This electric poultry netting is far from foolproof. Even after numerous lectures and demonstrations, some of our experienced apprentices and farm managers never quite got the gist of how an electric fence really works. Many times we would find improper knots in the fencing. Many of the ways they tried to secure the fence would short in out completely! You will save yourself a lot of time, trouble, and expense by understanding how electric fence really works. We feel this is so important that we have dedicated an entire chapter to it. This chapter is entitled: What You Don't Know About Electric Fence Can Be Shocking!

Just before dark every evening, one of us will go down and check the pens just as the birds are ready to settle down for the night. Some birds don't go back inside their shelter, especially at first. We gently herd them back inside, shuffling them along and sometimes picking them up and putting them inside. Finally, with everyone in, we close the doors. After a few nights, almost all of the birds know to go inside as part of their routine. We let them back out the next morning when we do chores. They usually come bounding out to meet us and be the first to catch that proverbial morning worm.

It is true that birds are early to bed and early to rise. If your birds have access to the outside they will go out and start foraging predawn at the first hint of light. We are not such early risers, so we don't let them out of their shelter until 8 am or so when we do chores. They lose a couple of hours of grazing time but that's not a problem. They will make up for it later in the day.

Also, when poultry, as with most animals, first wake up in the morning they poop. We want to capture that manure in the deep bedding inside the shelter so we can add it to our compost piles. The field outside the shelter gets enough traffic and fertilizer during the day to support the grass and legumes pasture.

If they are outside during the day time and it starts raining they will go back inside the shelter to keep from getting wet and cold. But if they bed down at night outside they won't get up and go inside if it starts raining. They'll just lie there and get chilled to the bone. They are night blind and can't see and will not move. We lost about 30 birds one summer night in a rain storm. The young birds had bedded down under the drip edge of the mini barn. They got so soaked and chilled that many of them died. This was a preventable loss!

Field Waterers

While the poultry are in the brooder, we use low-pressure bell drinkers that can run off a garden hose. They need a pressure regulator to reduce the pressure from the garden hose and feed the water through a 1/4-inch plastic tube. The drinkers cost about $15 each and the regulator about $12. One regulator can handle up to 10 bell drinkers.

Field Waterer

This Sweet Grass hen (Glenda) isn't really thirsty, she just wanted to be in the picture and is always available to pose for photo opportunities.

Rubber feed tub. If double sided, drill holes in side of tub to secure water tank float.

Elbow connector makes garden hose easier to connect to float and improves water flow.

Garden Hose

Flow regulator lets you keep the water pressure lower which helps prevent floods if the tub overturns. The "Y" connector lets you attach to another field waterer. Position the tub so that the float-end is downhill, otherwise the water level doesn't get high enough to raise the float and shut off the water flow. As you can see from the water level higher at the float and lower on the other side, this photo was taken on a steep hill side.

When we move the birds from the brooder to the field we give them a Fortex rubber tub and a stock-tank float-valve hooked to a garden hose. The rubber tub is the kind made to flex when frozen or stepped on so they are hard to break. They are about 4-inches deep, and will provide plenty of water for 200 to 400 birds. Fortex is one brand of these tubs, and are available at many feed stores.

Parts of a field waterer:

- Rubber tub (bucket - about $7)
- Stock tank float (preferably metal - about $15)
- Elbow connector (preferably brass - about $3)
- Water flow regulator and/or "Y" connector (about $5)
- Garden hoses, preferably contractor's grade

Ideally, your field water system has frost-free hydrants connected by underground PVC piping. This saves destroying good garden hoses with your bush hog when mowing. We've mowed a lot of hoses.

The entire field waterer setup, not including garden hoses or PVC pipes and frost-free hydrants, will cost around $30. We found the plastic connectors, while cheaper, didn't stand up to field use and all the moving around we did with the waterers. The metal or brass connectors are more expensive, but should serve you several seasons. Caution – don't let this system freeze. Even though it won't harm the tub, the connecting parts will be subject to breaking. If you expect a freeze overnight then disconnect and drain the hoses.

When the birds are smaller (2 to 3 weeks old), we put a poultry net screen across the top of the bucket to keep the little chicks from falling in the water, or if they do they can get out easier. With the shallow buckets, we have not had any chicks drown.

Every day when we do chores, we quickly rinse out the buckets to remove algae, wasted feed, dirt, and manure. We also move them a foot or two which changes the birds traffic patterns and does not wear the ground down too much.

The field waterers are useful for other livestock and pets as well. We also use this system on our deck and in the yard as a three-season waterer for our integrated pest control management swat team (dogs and cats).

Feed Troughs & Range Rafts

While the poultry are in the brooder, we use "baby brooder size" feed troughs that are available from poultry supply houses.

Range Raft Feeder

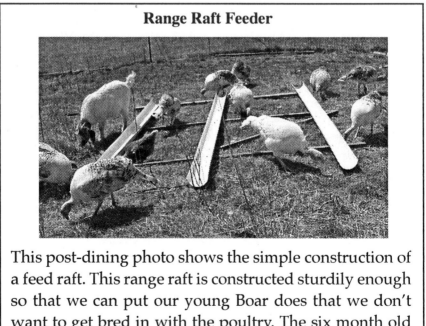

This post-dining photo shows the simple construction of a feed raft. This range raft is constructed sturdily enough so that we can put our young Boar does that we don't want to get bred in with the poultry. The six month old kids can use the extra feed for their growth. If you want to keep the goats away from the feed, you can make a simple creeper feeder by using a upside-down gate so that the turkeys and chickens can get under the gate, but the goats can't.

Once we move them to the field, we use trough feeders made from 6-inch PVC pipes. We use Schedule 40 PVC pipe, which is low cost and light weight. We use a chalk line and an electric circular saw or table saw to cut the pipes in half lenghwise. This gives us two feeders that are 10-feet long. We secure the PVC feeders to a wood frame using screws. We use 2x2" or 2x4" boards. Washers with the screws help keep the feeders from cracking or breaking loose from the frame.

Each feeder will hold five gallons of feed and cost about four dollars. It is very quick and easy to walk along the trough with the bucket lip resting on the trough and pour the feed in. Very little spills this way, and you can gently brush the poultry out of your way as you proceed.

Layaway Shelter Built on a Trailer

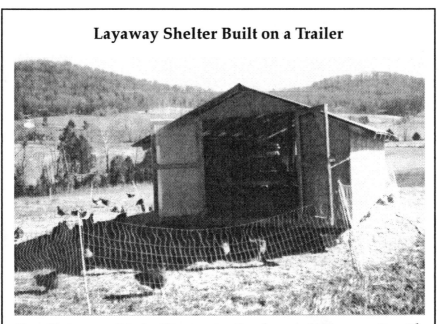

Our "layaway" is built on a trailer frame. We move it with a tractor to new pasture about every 2 weeks or whenever the ground shows too much wear. Note the doors open wide enough so that we can load and unload bedding with our front-end loader.

For our turkey range feeders we use Schedule 80 PVC pipe that is 20-feet long. The Schedule 80 pipe is much more expensive, but heavier and less likely for the turkeys to turn it over and spill the feed. Turkeys are much more aggressive at the feed trough than chickens. It's best to not let them run out of feed at all, that way they don't get frantic and charge you when you show up with feed.

We leave the ends of the troughs open so that rainwater can simply run out on the ground. The chickens and turkeys will happily eat any feed that gets wet or spilled.

We use 2x2-inch wooden skids that are 8-feet long under four of these troughs to make a "feed raft". With a rope handle we can pull the raft forward about 10 feet each day, then refill it

Tim Shell's Layaway with Roll-away Nest Boxes

Here is co-author Patricia Foreman admiring Tim Shell's very clever layaway design. He has put roll-away nest boxes inside, a walkway to collect the eggs, and space for his goats on the other end. The goats can't mingle with the chickens. He surrounds this with electric poultry netting and moves it with his tractor.

Here is Tim Shell collecting eggs with his daughter, Mary Frances, inside his layaway. Note how the eggs roll away from the hens nest box and into a collection isle. This really helps to keep eggs clean and makes collecting easier. Tim incubates these eggs to produce pasture peeps.

with feed. The raft will hold four 5-gallon buckets of feed, which is enough for about 400 small layers, or 300 broilers, or 200 turkeys.

Layaway Shelters for Laying Hens

Day Ranging layers makes a lot of sense. Laying hens are harder on pasture than either turkeys or broilers. Hens will scratch the soil and even dig small holes for dust bathing. They are more selective in their diet, and will pick out the more succulent morsels in the pasture, leaving behind the hardier grasses such as fescue, which they find unpalatable.

We use a layaway, which is a 12x21-foot hen house on wheels to hold our layer flock of about 250 hybrid brown egg layers.

There are many models and names for what we call "layaway". Polyface Farm calls theirs an "egg-mobile". Another model for a small back yard flock is a kit called a Hen-Spa.

Whatever you want to call your mobile hen house, it offers many advantages for grazing hens on pasture or in gardens to improve both the soil and the quality-of-life for the hens. The common elements are:

- A portable structure with a floor, often on wheels.
- Nest boxes inside.
- Portable electric poultry netting.
- Field waters and feeders.

Our layaway has a plywood floor covered with bedding so we can capture the manure for our compost. The sides are plywood, and the ends and doors are covered with poultry netting. In the winter we cover the ends with 6-mil plastic so we can keep the hens in the layaway year round. We move the layaway as needed to give them fresh pasture.

This layaway is called a Henspa designed by Egganic Industries. It comes as a kit and is good for about 12 hens. It is moved by hand using leverage and the wheels. For more information call 800-822-5866.

These layaways are built on hay wagons and were archi-
tectually designed for the hens' comfort, climate control
and easy egg collection. These are used on Sunnyside
Organic Farm, owned by Maggie and David Cole. David
is the good looking guy on the left.

Inside the Sunnyside layaways, the hens have slanted roosts
on the left, pull-out feed troughs in the middle, and nest
boxs on the right. The nest boxes are accessible from the
outside for easy collection. The floors have wire mesh under
the roosts that let the manure fall to the pasture below. The
open mesh assists in ventillation and cooling.

Two lengths of poultry netting will enclose about 7,000 square feet, which is enough pasture to last the hens one to two weeks, then we move them to a new spot. During the course of a year we might revisit a spot two or three times, with several months of lag time between visits.

We rotate the beef steers ahead of the layers. The beef cattle will mow the grass for the birds, and the birds scratch through the manure piles left by the beef cattle. This spreads the manure over a larger area, reducing or even eliminating typical "repugnance zones" when manure piles stay intact. The poultry also pick out any undigested grains, and any parasite larvae in the manure pats. We don't keep the beef steers in the same pen as the hens because the steers will rub against the layaway as a scratching post and will eat all the hens' feed.

We plan on 1/4 pound of feed per laying hen per day on average. When we gather eggs in the evening we grab the pull rope and move the feed raft forward a few feet. We also empty the water pan and rinse it and move it forward a few feet.

We provide water for the layers with the same field waterers we use for the broilers and turkeys. One bucket holds plenty of water for up to 300 hens.

Keep the feeder and water fountain within 100 to 200 feet of the layaway. If you put it any further than that the hens will slow down egg production because they have to travel too far for water and feed, then back to the layaway to get on a nest. They will range much farther than that on their own, but usually only after they've laid their egg for the day. By moving the layaway and feeder and water in a uniform pattern you can "sweep" the field, laying down a good dose of fertility and getting the sod scratched up, but not enough to hurt it. Baring the soil in places will help the clovers re-establish, so you will have a better pasture sward next year.

We use standard sized nest boxes with roosting bars. In the late afternoon when we pick the eggs, we put the roost bars up so the hens can't get back in the nest and roost there all night and poop all over the egg.

In the morning when we do chores, we put the roost bars back down which allows the hens to get into the nests to lay their eggs. We use synthetic nest box liners that are easy to clean and don't promote disease or pests such as mites. We take the liners out periodically and pressure wash them.

Nest box liners are inexpensive, about $2.00 each, and will last for many seasons. Because of the perforations in the plastic liners, the air circulates around chicken droppings and dries themvery quickly. Then to clean the nest, we just pull out the liner and slap it against a roost to dislodge dried manure, then reinstall it. These plastic nest liners will virtually eliminate dirty eggs that are caused by a dirty nests.

We have tried planer shavings for lining nests, but it gets raked out of nest easily. Hay or straw will stay in the nest, but it is hard to keep clean and harbors mites and dust. Aspergillosis, a respiratory illness caused by dust and molds in spoiled hay or straw, is a big concern for poultry growers. That's why we use plastic liners instead.

We run an extension cord out to the field in the winter time to power a light so the hens will lay more eggs. If you use a florescent light make sure it has a cold start ballast. A 4-foot fixture with two 40-watt florescent light bulbs will provide enough light for 300 hens. Have it set on a timer so it provides 14 to 16 hours of light per day. Set it to come on two hours before sun up, then shut off during the day, then come on again at dusk and light them well into the evening.

Have a second, much dimmer night-light type of bulb to come on at the same time the big light goes off at night. This gives the hens enough light to find their way to the roosts and bed down for the rest of the night. Otherwise the sudden darkness will terrify them causing unnecessary stress and inability to get settled onto their roosts. The less bright light does not need to be on all night.

The layaway has roosts which are 2x4's on edge, set 16-inches apart, and allowing 6-inches of roost space per hybrid hen. Our dual purpose (broad-bottomed) hens are usually much larger than the hybrid hens so we allow at least 8-inches of roost space.

Not everything worked as planned. This early version of a 300 turkey shelter was a failed covered-wagon model. Yes, the past-tense, as in "was". It had flawed design features. First, even though there were roosts across the bed of the hay-wagon frame, the turkeys would settle down only around the outside rim forcing the others to roost underneath or out in the open. And worse yet, the outside rim is also the drip edge so the birds got soaked in a rain. Second, the roofing ripped off in a high wind storm. We dismantled it and recycled the lumber for other projects.

Our layaway has 144 lineal feet of roost space, enough for 250 small hybrid hens, or 200 dual purpose hens. The layaway is 252 square feet, which is large enough for 250 small hybrid hens or 200 dual purpose hens.

What Kind of Layers?

We have tried dual purpose meat and egg hens. Not surprising, we found they don't lay as many eggs as the hybrids, they eat more feed, and they take up more shelter space. They average about 6 pounds live weight. At the end of their egg laying career, we can process them and sell them as stew hens for about $6 each. However, we don't have a big market for stew hens, and sell most of them as pet food.

We have found that the dual purpose birds aren't really very good for either egg laying or meat production. They only lay about 200 eggs per cycle and they don't have the wide breast that we prefer in meat birds.

Often we give the dual purpose hens away to people who just want a few eggs, or who want to eat the birds. If we can't find a home for the dual purpose birds within a reasonable length of time, we kill and compost them.

In recent years we have decided to use smaller hybrid hens rather than dual purpose hens. The hybrids lay more eggs, don't eat as much, and because of their smaller size they don't take up as much space in the shelter. At the end of their egg laying cycle they are of very little meat value because they are so scrawny. We find they aren't worth processing for the little bit of meat you do get. We either give them away, or kill and compost them.

We have tried several strains of egg laying hybrids such as Production Red, Production Barred Rock, and Golden Comet. These are all brown egg layers, and will do well in a layaway flock. We prefer the Golden Comets because they are docile

This historical photo is yet another example of what did not work for us — or for our birds. It shows two of our heavy 3'x10'x12' bottomless pasture pens with doors (in back), a 3' PVC turkey shelter (left foregound), and a 2' chicken shelter (right foreground). We were experimenting with these three types of shelters by tracking weight gain.

Notice that the birds in the back pens are free to come and go. What we found was that some birds in the PVC shelters (foreground) were dying because of the heat and overcrowding. They were dirty, manure covered, and not happy campers. But none of our "free" birds died. They were clean and content. Duh! A blinding glimpse of the obvious (BGO)!

We abandoned the experiment and raised the corners of the PVC shelters to let our birds out — the death rate dropped to zero. This became our motivation to explore possibilities of the Day Range system using shelters with bottoms combined with electric poultry netting, range-raft feeders, and field waters. Hence, this book.

These are the PVC pens shown earlier after just one winter. The 2 foot high pen on the right was picked up by the wind like a box kite and smashed apart. The 3 foot pen on the left has cracked posts and is virtually immobile. We don't recommend PVC pipes for pasture pens.

and friendly and we like the feather coloring of a reddish background with white tips.

Day Range Broilers

The Day ranging system really shines when raising meat birds. Our goals are to provide sturdy housing to protect them from weather and predators, and ensure plenty of pasture for foraging and getting exercise and sunshine.

We keep the chicks in the electric brooder for two weeks. Depending on the weather, we might keep a heat lamp on them another week or two in the floored mini-barns. Our goal is to keep all our birds unstressed — this includes keeping them warm and dry.

After they have hardened off and are used to being inside the mini-barn, we let them out on pasture inside the poultry netting. Here's a tip: By letting them get used to being and sleeping in the mini-barn for a few days to a week, the birds will be more inclined to go inside by themselves at night.

Unlike layers, broilers sometimes don't go back inside the mini-barn to roost at night. Often they will simply lie down wherever they happen to be when it gets dark. That's not a problem if you have a good electric poultry netting around them to keep out predators and if the weather doesn't dump a cold, hard rain on them in the middle of the night. Once they have bedded down and it gets dark, they will not move on their own, so a cold rain will soak them and chill them and many will die.

We find it necessary to check the pens each evening and chase the stragglers back inside the shelter. If we are not having a problem with owls killing the broilers during the night, we will leave the shelter door open so that at daybreak the broilers will get up and go out to pasture. However, if we are losing some to owls every night, we then chase the broilers back in and close the door so they are safe overnight.

By using the electric poultry nets and closing the birds in at night, we have nearly eliminated predator losses in our broiler flocks. And by hardening them off properly from the brooder and putting them back in the shelter at night so they don't get rained on, we have virtually eliminated death losses due to weather.

A big advantage of the mini-barn or hoop house being in the middle of the paddock is that the birds can rotate around the shelter as the shade moves during the day. Many of the birds will go back inside the shelter for shade. Others will nestle down around the outside of the shelter in the shade, and slowly move with the shade as it circles the building.

We move the broilers around inside the paddock by moving the feed troughs and water fountains. That spreads traffic and wear and tear on the soil.

Another big advantage to the hoop house are the wide doors at each end. They make it easy for the birds to come and go as they please, and the wider threshold spreads the foot traffic at the doorways. Still, if you leave the shelter in place for 5 weeks, they will probably wear out the grass at the entry. That's okay, it will grow back as soon as you harvest the birds and move the shelter away to the next spot.

Day Range Turkeys

We figured we could raise 500 turkeys on 5 acres. We would fence it in with 2'x4'x4' welded wire and electric predator wire, plus poultry netting on the ground around the perimeter to

This is one of our mini-barns that multi-function so well for homesteads. We have winterized this one with plywood sides and plastic ends. We located this barn close to our house and in our kitchen garden for two reasons: First, to make taking care of the laying hens easier during the winter. Second, to have the hens fertilize and clean our garden beds.

There are nest boxes inside. Notice we leveled it with blocks because it would be stationary for several months. We put chicken wire skirts to keep the hens from nesting under the floor. We also used the mini-barns for our goats to do rotational grazing.

keep predators from digging under. The fencing would cost about $1,500 not including labor.

Then, we would need roosting shelters—either 10 shelters at 8x16 feet or 5 shelters at 16x16 feet. Since it might be difficult to move the larger units we considered temporary pole sheds of 16 x 16 feet. These would be near the center of the 5 acre paddock and we'd use straw or planer shavings from the Virginia Horse Center to keep the ground from muddying around it, and for bedding under the roosts.

If the roosts were made so they can come out easily then we could use the tractor front-end loader to remove the bedding at the end of the season.

This 5 acre paddock would then be ideal for turkeys for the first two years. Then, for stocker beef and ultimately dairy cattle. Each year the grass gets richer and richer and eventually the forage will be very valuable as a site for grass-fed dairy. One rotation could even be for high value, high yield vegetables and fruits.

The advantage of the stationary shelter is that we can build it on fairly uneven ground and level it. And, since we never have to move it, we can orient it to a proper southeast front and leave it in place for years. In the off years between turkey flocks, we can use the shelter for cattle or horses in the same manner. Just take out the roosts and bed it down.

As a general rule you can raise 100 turkeys per acre. They are only on the pasture for 8 to 12 weeks before Thanksgiving, but that's in the growth months before killing frosts arrive. Irrigated pastures may be useful too.

Shelters, Mini-barns, & Hoop-houses

Over the years we have invested a lot of money researching numerous shelter designs. We found that there are several designs that are worthwhile, depending on what you are trying to accomplish.

If you are only growing birds in the milder seasons, then you can get by with a lot less shelter than if you are growing them in the winter, too. Of course, everything is relative, with the northern states being a lot cooler in the summer than in the south. It is entirely possible to grow broilers and turkeys and layers year round in most climate zones, if you can provide the kind of shelter they need.

Shelter in the south, especially in the summer time, must be designed for maximum air flow and shade. In rainy zones it's necessary to provide rain shelter, yet with air flow so that molds and mildew don't become a problem.

When we first started out with chicken tractors over a decade ago, we were only interested in growing a few broilers for our freezer and a few to sell. In that case a typical 4 x 8 foot chicken tractor was large enough for our needs.

Chicken Tractor 4x8 Pens. On our farm, we have discontinued using the 4x8-foot chicken tractor style. For our purposes it is too expensive for the number of birds it will hold, and will not provide adequate shelter except during mild summer months. These pens did not have floors, although we could retrofit them with bottoms. This model of chicken tractor is still good for the summer flock in a kitchen garden. We still recommend adding pop holes (bird sized doors) to let the birds out to graze more in addition to having some sort of protective fencing.

On a larger scale however, we found the time it takes to feed and water the chicken tractor flock (only about 20 hens) is the same that it takes to do our layaway flock of 250 birds. The 4x8 pens are still useful for folks who are growing a limited number of birds.

10x12 Pasture Pens. We have also discontinued using pasture pens that measured 10 x 12 feet and were 2 feet high. Typically, three fourths of the roof was covered with some form of sheet metal roofing. Chicken wire covered the remaining quarter of the roof, and all four sides. We found these pens to be very difficult to move on our rough pastures. Unless we were very careful, we wound up crushing birds, hurting our backs, or doing both, even when we used a dolly and/or attached wheels.

The pens were too expensive to build for the small number of birds they hold. The low metal roof heated up way beyond reason on hot days and killed birds. The sun slanted in under the open sides and the birds crowed into a corner trying to find shade, which resulted in them piling up and over heating. They didn't have a lot of grass to eat because they were so close together, and by the time they all had a bowel movement most of the grass was coated with manure, or they had bedded down on the grass and crushed it. Birds raised in this manner were of uneven sizes, some scrawny and some fat, with a lot of scratch marks on their backs from piling and lots of broken wings and legs that got trapped under the pen edges during moves.

PVC 10x12 Pasture Pens. A modification of the typical 10x12 foot pasture pen is to use PVC pipes for the framing members. This works reasonably well, but the PVC breaks under pressure so the pens are not very long lived. They are also very expensive – probably twice the cost of a wooden pen.

One way to improve on the 10x12 foot pasture pen is to cover the entire roof, put wings on all four sides, raise the roof to 3-feet or even 4-feet, and put wheels on it.

The wings on all four sides close down in the evening to enclose the birds. The wings open up during the day to release the birds, and form a more shaded area for the birds to get under with more roof protection if it rains.

Raising the roof an extra foot or so allows more air flow so the birds won't overheat on a hot sunny day. Covering the entire roof allows more shade. Putting wheels under the pen makes it very easy to move by hand on level ground, or with a tractor or pickup truck on hills. Then use the electric poultry netting to contain the poultry inside the paddock. Photos of the PVC pens are shown earlier in this chapter.

Heavy Duty Multi-Purpose Chicken Tractors. We built 4 heavy duty chicken tractors that measure 8-feet wide, 16-feet long, and 4-feet high. We also use them as pig tractors and as goat tractors. We even used one of them to provide field shelter for a pair of beef steers one winter. We also used them as turkey

Hoop House Shelter

Loading fresh wood-shaving bedding into the hoop house shelter. We will collect the nitrogen-rich, manure-laden bedding later to make compost for our gardens. Notice the doors are wide enough for our front-end loader to dump the bedding in — we like labor to be non-intensive. So far the hoop house is our favorite Day Range shelter. But the perfect one has yet to be developed. Perhaps you will design it. If not you — then who? Someone needs to carry on...

shelters, but found the turkeys were apt to fly up to the roof and roost rather than go inside. We built these pens heavy duty because of the terrain and variety of uses we expected. Even though they are heavy duty, we had one collapse one year when about 50 full grown turkeys tried to roost on it.

We built the 8x16 chicken tractors on lawn mower wheels to make them easier to move. Here's how we attached lawn mower wheels to the frame. We first bored a 1/2-inch hole through the bottom frame member. Then, we inserted a 1x2-inch by 6-inch bolt through a lawn mower wheel, then through the frame member, then through another lawn mower wheel, and then locked it in place with two 1/2-inch nuts.

This way the 6-inch bolt acts as an "axle" and keeps the wheels from rubbing against the frame and coming off. On a 16-foot long chicken tractor we put the wheels about 1-foot from each end and bored the axle holes so that the wheels would hold the frame about 2 inches off the ground. This gives enough clearance to roll the pen along, but not so much clearance that the birds could get out, or predators could get in.

Of course, once you've gone to the effort of improving a pasture pen to where it is most useful, you will be only a couple of steps away from designing and building a mini-barn which will be even more useful.

Mini-Barns. Pat also calls these, "barn extenders" because they are an extension of our large pole barn. We first built mini-barns that were 8-feet wide and 16-feet long, with 3-foot sides and a pitched roof high enough to allow us to walk inside. This is important when collecting eggs or servicing hens during a winter storm when we keep the birds inside.

The sides and ends are enclosed with chicken wire and the roof is galvanized tin. These mini-barns also have a pressure treated plywood floor that is covered with planer shavings as bedding.

The mini-barn rests on pressure treated skids, making it easy to move with our farm tractor.

These mini-barns are expensive, but we built them heavy duty so they can be used to shelter beef calves, goats, pigs, and also for year round use as poultry houses. The other reason we built them so heavy duty is so they will stand up to the rough moves over our uneven terrain, and they don't become kites in windstorms.

Some bottomless shelters become box-kites in high winds.

Hoop House Shelters. We found we still needed a better shelter design than the mini-barns for larger production. This is because the mini-barns are (1) relatively expensive and (2) still don't hold enough broilers. Mini-barns hold approximately 128 birds allowing 1 square foot each.

We tried making a shelter out of PVC hoops, and gave up on that because the PVC plastic tends to break in cold weather. PVC also can't stand up to the wind and snow loads that we have in our area. After seeing that Tim Shell's PVC house had collapsed three times, we chose to use heavy duty metal greenhouse bows.

The metal greenhouse bows are more expensive than EMT or PVC, but we find they are much sturdier and already bent to shape. This makes them much easier to work with. They are already engineered to withstand the snow and wind loads common in our area.

We built our first hoop house shelter using 14-foot wide bows, which were left over from a greenhouse we had built on our farm. The bows are 1-1/4 inch in diameter and galvanized, and will probably last for a lifetime. We attached the bows

House Trailer as a Day Range Shelter

Andy bought this pre-1976 trailer for a song at a farm auction thinking we could house apprentices. Because it was so old, it did not meet code and we wouldn't legally use it for housing. So we gutted it and turned it into a turkey day-care brooder. As the turkeys grew, we used it as a turkey shelter that put our failed covered-wagon model to shame. We trained the turkeys to roost inside at night, and with the help of electric poultry fencing, protected them from any harm both day and night. Usually these trailers end up at the dump – and probably rightly so... The next year, we used the same spot to grow the BEST melons.

Turkey baby day-care inside an old house trailer.

on 5-foot centers to pressure treated skids. Then we laid in a pressure treated plywood floor and covered the bows with a 22-mil woven poly cover.

This gives us a 14x16-foot shelter, measuring 224 square feet, and perfectly sized to raise 200 to 250 broilers per batch. The hoop house can also be used for layers, turkeys, and goats. We don't put roosts inside for the broilers, but we would for layers or turkeys.

The hoop house shelter is about half the weight of the 128 square foot mini-barns, by about one half, yet they are heavy enough and aerodynamic enough to withstand wind and snow loads we have here in Zone 6 of the Shenandoah Valley. We move it with the tractor. Each end has a pair of doors that open up to allow the birds in and out. It is wide enough so we can put our tractor bucket inside to load or unload shavings for bedding. We use 2x4 thresholds at the doors to hold the bedding in place.

We always sit the hoop house to take advantage of the prevailing southwesterly summer breezes. Even on the hottest of days here in the valley, we still get enough air movement through the shelter to keep it pleasant. The air temperature inside the hoop house is always lower than the day time temperatures, and at night the woven plastic cover helps keep in enough heat from the birds so they don't get chilled.

A driving wind will blow rain partly into the shelter at night, but because they are on dry bedding and a floor that keeps them off the cold ground, the birds are less likely to get hypothermic.

The ends of the hoop house are framed with 2x4s covered with poultry wire. We can easily enclose the ends with 6-mil plastic to provide more shelter to the birds and extend our season in the spring and fall. With covered ends, we can also use the hoop house as a floor brooder by just adding some heat lamps.

We use a floor in the hoop house for three reasons.

- First and most important, we need and want to protect the birds from the cold, wet ground at night during all seasons.

- Second, we like to capture the manure in the bedding for our compost pile.

- Third, the plywood floor acts as a diaphragm, supporting the sides and ends when we are moving the rig to a new spot. This keeps everything aligned so that the building doesn't pull out of square as we are moving it.

Yes, the plywood floor adds considerable expense to the unit, but the payback comes fast with healthier birds, easier chores, and longer lasting structures.

We usually leave the shelter in one spot for the life of that flock, which is usually 4 or 5 weeks on pasture. Moving poultry is very stressful to the birds so we don't do it unless absolutely necessary. We move the shelter to the next spot for the next flock.

Also, by not moving the flock, we never crush or injure a bird during moving. The grass under the hoop house will go dormant from lack of sunlight, but will start to regrow as soon as we move the shelter and give the clovers a chance to come in. If there is any damage to the pasture where the door entries were we can throw on some grass seed and cover it with some of the bedding, which will help keep the seed moist until it germinates.

By the following year, it is all but impossible to tell where the door entries were, unless the grass seed we are using is quite a bit different from the other pasture grasses. We are delighted to find that native clovers are reappearing in our pastures.

If you should need to move a flock with any of the Day Range floored barns, it is easy. Just keep the birds in from the night before, or chase them all inside and close the door. Hook the the shelter to the tractor and pull it where you want it. The birds are totally safe.

Catching Birds for Processing

With Day Range it is easy to catch the birds for processing. You can do it inside or outside, the mini-barns, hoop houses, or other designs.

With the birds inside the shelter after dark, we can simply close the door, put the crates inside the shelter, and gently catch the birds. We then put the crates back outside on the grass overnight. We do this so the bird poop is spread back on the land instead of on each other as would happen if we were to stack the crates up at night. Of course, if it's raining outside, we leave the crates inside the hoop house overnight to protect the birds from the rain.

Another way to catch them is to lock the birds out of the shelter in the evening so they will bed down against the building. Then after dark, move in and surround the birds with the plastic crates, and quietly and gently pick them up two or three at a time and put them in the crates. Be respectful not to scare them. In the morning, bring the trailer down and pick up the crates and deliver them to the processing room.

The only time this night catching scheme doesn't work well is when we don't wait until full dark, which can be 9 p.m. around the summer solstice, or with a full moon. If it is too early, a few of the birds can see well enough to run and then we have to chase them.

The way to avoid this is to make up a simple "trap fence" that can be set up around the nestled birds. The trap fence is 2-feet high, and made of poultry wire, so they can't get through or jump over it. This makes it very easy to catch them.

Here Andy is using circled transport crates to form a mini-corral to catch chickens for processing. This is easier to do at night because chickens and turkeys are night-blind.

Turkey Shelter Systems

The largest shelter you can build practically for day ranging and can still drag around is perhaps 12x 16-feet. But we have found the 8x16-foot size is easier, and requires smaller framing members and less bracing. Anything larger than 12x16-feet is apt to collapse as you try to pull it with the tractor. The steep roof of a twelve-pitch building will be enough to keep the birds from roosting on the roof. However, the flat roof is much easier and less expensive to build. Birds will roost on the flat roof, but that's okay. If you don't want them to roost, you can build an anti-fly device using lighter framing materials and poultry mesh.

With the sides of the turkey tractor open they will come and go at will and will be much more likely to go under shelter in bad weather. Smaller turkeys are apt to stay out in the rain and pile up for warmth and possibly smother each other.

For 100 turkeys growing up to 20 pounds live weight, you will want to give them each about 18-inces of roost space. You can do this in two shelters that are 8x16-feet and are on wheels so you can move them easily with a draw rope. That gives them 148 lineal feet of roost space, or about 18-inches each. It will be rare, however, that they will all go into the shelter. Even during a bad storm, some of them will just roost down on the ground and hunker there through the night.

Pasture Renovations, Hillsides, Rough Terrain, Crop Rotations, and Multi-Species Grazing

Our farm has rolling hills that are not well suited to conventional agriculture. The infertile soils are highly susceptible to erosion. The region is ideally suited to graze, and for the twelve years prior to our purchase, our farm had been a sheep farm. None of the land had been fertilized or limed for at least two decades, and as a result the topsoil is pretty well devoid of nutrients.

Given the poor state of pasture and soil on our land we could be tempted to plow everything down and start all over again in order to establish good pastures.

If we follow the USDA Cooperative Extension recommendations for establishing good pasture we would use herbicides to kill the existing growth then use a "sod-planter" to sow new seeds. To improve pasture "organically", we would need to mow the existing grass then plow it down, disc harrow, fertilize, lime, and reseed – all this at a cost of about $200 per acre or more. To do our 30 acres of pasture and hay fields would cost us about $6,000. Our yields from that investment in pasture renovation would then amount to about $300 net per acre per year in stocker beef gains or hay production. To optimize hay or beef returns, we would need to fertilize and lime the pastures at regular intervals.

It's better for us to invest that $6,000 in an income earning farm enterprise such as broilers, layers, or turkeys. We fence in an area, turn some poultry on it, give them plenty of feed and water, and let them graze off whatever they can find in the worn-out pasture. Along the way, we can earn an income from the birds while they improve our pasture. Here's an example of how that can work. If we have 100 turkeys per acre and make $20 net income per turkey, then we earn $4,000 per acre instead of having to spend $200 per acre as the USDA recommends.

Most of the information we have to go on when designing poultry systems on pasture comes from old books from the 1920's to 1950's, before there even was a confinement broiler industry, and when almost all turkeys were grown "on range". According to USDA research from that era, each pound of feed given to poultry on pasture returns 1.1 to 1.5 pounds of manure, of which 75% is water.

Turkeys and chickens are not especially good at extracting available nutrients from their feed. The resulting manure is quite high in nitrogen, phosphorus, and potassium as well as many trace elements. This is especially true at our farm since we feed our flocks a wonderfully rich diet containing kelp, diatomaceous earth, aragonite, and Fertrell's Nutri-balancer. This is in addition to organically grown wheat, oats, roasted soybeans, and corn.

Turkeys on pasture can put down about 60 pounds of manure each, or 3 tons per acre per 100 turkeys. Nutrient variability is extreme, but we figure the manure has a 10-10-10 value of nitrogen, phosphorous, and potassium, plus a wide range of micronutrients. Therefore, we're putting down easily 240 pounds of nitrogen per acre on a dry weight basis. This would be way too high for the following crop except that the manure is placed on the sward over a 8 to 12 week period, thus giving the soil microbes a chance to metabolize the manure slowly as it is applied.

As is the case with many organic fertilizers, only about 50% of the available nutrients are released per year. Only if you leave the turkeys in one place too long will the manure overload become excessive. This occurs mostly in their roosting area since they will deposit a third of their daily manure on that small area.

That's why it's important to use the fence, feeder, fountain, and shelter as the brakes, steering wheel, accelerator, and engine of the feeding system. The perimeter fence keeps the birds in and the predators out. We move the shelter weekly, and the feeders and fountains daily, so the poultry are always traveling, grazing, resting, and pooping in a different part of their paddock each day. Whatever our objectives are, we can accomplish them simply by moving one or all of the furnishings that make up their pen.

The year following the poultry rotation, we will have a richly fertilized sward. If we manage the sward carefully to encourage legumes, we can have a forage very high in protein, enough so that we could pasture lactating dairy cows profitably.

At our farm we don't have a high-value livestock enterprise such as dairy cows to follow the turkeys, so we rely on the stocker beef to soak up the extra grass. Increases in earthworm population are magnificent. Killing fescue allows clovers and other legume and grass seeds that are dormant in the ground to germinate.

Multi-cropping Poultry with Vegetables & Hay

Whenever possible we rotate our turkeys, layers, or broilers through our market garden, berry patches, and hay crops as a way to fertilize them.

This is especially appropriate with turkeys, because they are on pasture at a time of year when other crops are finished. We grow our largest number of turkeys for Thanksgiving, and our customers want turkeys ranging in size from 12 pounds to 20 pounds dressed weight. If we start our turkeys during the third week of July, they will be just the right size for Thanksgiving. The small hens will be 12 pounds dressed weight, and the largest toms will be 20 pounds or better. The turkeys gain weight at an average of 1.25 pounds per week, live weight. A 16 week old turkey will then weigh about 20 pounds live weight, and will dress out to about 15 pounds dressed weight.

If our poults hatch on July 21st, they will be in the brooder for 8 weeks, or until about September 21st. By that time, many of our market garden crops have finished for the year. Then we use the electric poultry netting to fence in part of a market garden, and turn the turkeys in there to hog down any crop residue, weeds, and grasses. In a matter of days, they can graze it right down to bare dirt. We then move them off that spot, and mulch it with old hay or straw for the winter. Next spring we have a wonderfully rich garden soil.

In normal turkey broiler management, the turkeys will only be on pasture for 8 to 12 weeks, in the late summer and fall. Plan your grazing rotations accordingly. Fence in an alfalfa field, for example, and let the turkeys graze it rather than take the last cutting.

On a farm such as ours, soil fertility is the same as having money in the bank. Whenever we want a good crop, we use the poultry to eat off the excess growth and to lay down a layer of manure for fertility. However, too much manure is a problem. around the major poultry production areas we are seeing an increase in phosphates because of excessive poultry manure being applied to the land.

States are now trying to regulate how much manure can be spread in any given watershed area, thus reducing the amount of pollution in our creeks, rivers, and bays. Farmers in Switzerland have long faced this problem. It is now mandated that Swiss farmers can't import more feed than their farm can utilize in manure. That's how we need to think of over feeding our individual paddocks. We want to put on lots of manure during the first years to get the ground back to state of fertility, but less manure in subsequent years so that over fertilizing, leaching, and contaminated farm runoff doesn't become a problem.

We also use our turkeys to graze an orchard. If we move them through fairly quickly they won't damage the trees. They will graze and forage through the orchard floor, and sometimes will browse the trees if they can reach the lower limbs. This is most noticeable in dwarf trees or young trees that haven't grown above the turkey browse line, yet. A pleasant benefit of having orchards is in using the tree prunnings to smoke the birds and the same with pigs. The fruit tree wood imparts a very pleasant flavor to the meat and poultry that is being smoke-cooked.

On-Farm Research

I believe that the pasture poultry movement has matured to a point where we can start looking for ways to fine tune our model to optimize yields. One way to do this is to undertake some on-farm research, using well thought out research models, and replicated trials. We need replicated trials so we can declare our results to be accurate. It's easy to give credit where it might not always be due. For example, here's what one grower did to improve weight gains:

1. Switched to organic grain.

2. Held chicks in the brooder for an extra week to harden them off.

3. Set up a new field pen system centered around a new hoop house design.

4. Switched to a different Cornish Cross broiler strain.

Now, the grower is satisfied with the weight gains, and brags about how "that organic grain is fantastic!" In reality, he hasn't a clue as to which of the 4 steps actually improved his weight gains. Most likely, all four steps were critically important.

Another article recently credited a magic potion of liquid probiotics with cutting mortality rates from 5-7% to less than 1%, without any other changes. And it only costs 1-cent per chick. This is the kind of report that begs to be challenged. If we prove, through 3 individual closely monitored tests, that chicks have a higher survivability with liquid probiotics, then we should all be adding it to the chick water.

In one on-farm research project, the grower was estimating each bucket of feed weighs 30 pounds. In our experience, each bucket of feed weighs anywhere from 25 to 35 pounds, depending on a LOT of variable factors. If we just _estimate_ weights then we can only _estimate_ results.

The problem with actually doing on-farm research and tracking the results is that none of us really have the time to do it. It all sounds so simple — just a few extra minutes each day to weigh the buckets of feed and mark it down on the clip board at each pen.

But when it's raining, or the pigs get out and disrupt chores, or you get distracted by any one of a million things that can go wrong on a farm, or there are several people sharing chores, the buckets get mixed up, and we forget to make the chart entries. Finally, at the end of the season, we just scrub the whole research project because the data is so unreliable.

But in the face of this, we still need to do on-farm research. At this point we simply don't have all the answers we need. One recent research project showed a wide range of nutritional differences between "free range" and "industry standard" meat and eggs. But this is only one study, and when we showed it to the poultry scientists at Virginia Tech they scoffed at it, and pointed out several ways in which the test results can be interpreted to prove a point.

We still believe the free range system gives better results than industry standards. We'd really like to have more data to back up our beliefs. We'd love to be able to show a dozen test results to the disbelievers at Virginia Tech and do a little scoffing back.

Most importantly, how can we tell our customers and potential customers for sure that our system is the best?

In summary, we'd like to challenge everyone reading this book to undertake an on-farm research project next year. Keep it small, keep it simple, and replicate it three times to test the results. Then crow about it so that others can learn from your experiences.

Chapter 8: Flock Management

In this chapter we discuss the ins and outs of different considerations of flock management. At various times during the year
we kept five different types of poultry on our farm. Each has
its particular management needs. The five flocks are:

1. Broiler Flocks (Meat Production)

2. Layer Flock Management (Egg Production)

3. Turkey Flock Management (Meat Production)

4. Turkey Breeder Flocks (Next Generation Production)

5. Broiler Breeder Flocks (Next Generation Production)

Don't worry, this is not complicated. There is a lot of information overlap in managing the different flocks. Once you know
how to care for one type the others are very similar.

1. Broiler Flock Management (Meat Production)

Once you have built your Day Range shelter and the temporary
paddock it's time to stock it with poultry. Each batch of broilers (or any poultry) of the same age should be separated from
broilers of a different age. If you put smaller birds in with larger
birds the smaller ones will get trampled on and pushed away
from the feeder by the larger birds. This results in the smaller
birds becoming scrawny and tough and the larger birds getting
fat. This is especially true of broilers and turkeys. Layers seem
to have more tolerance for mixing with different age groups.

Some growers have tried raising as many as 400 broilers in an
8x16-foot mini-barn, which totals 128 square feet of space. Our
experience indicates this is terribly overcrowded. We limit each
individual flock to 200 birds or less, and give them a mini-barn
shelter that measures 14x16-feet, or 224 square feet of space.
With this size flock the shelter does not have to be large and

unwieldy, and the birds don't have to compete as hard with each other for food or space.

If something should go wrong, such as disease, predators, heavy rain, or blazing sun, we don't risk losing as many birds in these smaller flocks.

Our plan is to move the broilers out of the brooder at the end of their 2nd week. We install them in the range shelter and leave them inside it for at least one more week. If the weather is cold or rainy or blazing sun then we leave them inside the shelter so they won't be harmed by the weather. This enables them to acclimatize to the outdoors gradually, without shocking their systems. This is very similar to hardening off greenhouse seedlings before placing them in the field.

At the end of the broiler chicks' 3rd or 4th week, we open the doors of the shelter and let them start free ranging inside the temporary paddock. By the end of the 7th or 8th week, we harvest them and move the shelter on to the next location in the field.

We always expect some trampling of grass and some bare spots to appear, especially at the doors to the shelter where there is the most concentrated traffic. When we finish the flock and move the shelter, we reseed the area that is worn, or let it regenerate naturally, depending on how extensive the damage.

The bedding inside the mini-barn is valuable for our compost piles, so we don't mind shoveling bedding at the end of each flock.

By using a longer brooder time and a more protective Day Range shelter, we can extend the pasturing season for poultry by at least a month in the spring and another month in the fall. This gives us more flocks each year, and enables us to spread our equipment costs over more birds. In some climate zones, it is even possible to raise poultry on pasture year round, by judicious planning and use of the Day Range shelters and pens. I know of at least one grower in South Carolina who is raising broilers year round.

To collect broilers for processing is very simple. On the evening before processing day just lock the broilers OUT of their shelter. As night falls, they will huddle against the sides of the shelter and settle down for the night. After dark, just slip in and surround them with your poultry crates. Gently and quietly pick up two or four broilers at a time and slide them into the top door of the poultry crates. Put 10 broilers per crate if they weigh less than 5 pounds live weight. If they are larger only put 8 broilers per crate. The crates measure 10-inches high, by 2-feet wide and 3-feet long. They are available from poultry supply houses.

After filling the crates with the number of chickens you wish to catch, leave the crates on the ground, inside the electrified poultry netting for the evening. The birds will sleep through the night and upon awakening the next morning will defecate, leaving their manure on the pasture. However, if rain threatens during the night, move the loaded crates into the mini-barn where the poultry will be protected.

In the morning, just collect the crates onto a trailer or pickup truck and carry them to your processing facility. Stack them in the shade in the area where you will process them.

Its best to withhold feed the evening before, so the digestive system will be empty for butchering. This eliminates most of the fecal matter and cuts down on the chance of contaminating the meat during processing.

2. Layer Flock Management (Egg Production)

Many small-scale poultry producers get their start by selling eggs from a Day Range flock. It is a fairly simple business to start, and in most areas of the country there is a ready market for the eggs. Care must be taken however, to sell the eggs at a price that is sufficiently high to guarantee a decent return on investment (see Chapter 3, Page 71, Marketing).

There are two types of egg-laying hens, "dual-purpose" and "hybrid layers". The "dual-purpose" are breeds such as the Australorp, Barred Rock, Rhode Island Red, and Orpingtons. These all lay a satisfactory number of eggs and their body size is large enough so that at the end of their laying cycle can be butchered for meat as "stew hens". The hybrid layers are bred for a smaller body frame and larger egg production.

You can start a flock either by buying pullets or layers from someone else, or by starting with day old chicks. We encourage you to start with day old chicks, because you can be somewhat assured that the birds will be disease and pest free when they arrive at your farm. If your farm is organic, then you have to start with day old chicks so you can maintain organic feed from day one.

Buy from one of the reputable hatcheries in your area. We recommend that you buy sex-link layer chicks. These are hybrid strains that are small framed yet lay a consistent number of eggs. Because of their smaller frame size they require less feed and less shelter space than the larger dual-purpose breeds. Our favorite is the Golden Comet, a small reddish-blonde bird that lays a good quantity of large brown eggs. They are docile, easy to keep, and will last for two years before needing to be replaced.

Don't ask me HOW, Don't ask me WHY and DON'T ask me to do it again!

A young hen's view and opinion of her first egg.

The one disadvantage to the small-framed hybrids is that at the end of their laying span they are too small to process and sell as "stew hens". However, over the course of their laying period, they will eat less feed and take up less space so you can put in more birds per shelter. With these attributes, and the fact that there really isn't much of a market for "stew hens" we find it just as easy to put down the hens and compost them, without going to the trouble of processing them for sale.

We provide each hen with 6 to 8 inches of roost space inside a "layaway", which is a hen house on wheels. We surround the layaway with electric poultry netting to form a yard for the hens. After they have grazed down the area and scratched up the soil a bit then we move them on to a new location.

We use galvanized nest boxes that have built in "roosting bars". The purpose of the bars is to close off the nests at night so the hens don't roost there and deposit manure in the nest. This helps keep the nests cleaner and the eggs cleaner. It also breaks up the "broody" cycle, when the hen wants to sit on the nest brooding the eggs, rather than laying more eggs.

Sometimes broody hens need to be separated from the flock for a few days to encourage them to start laying eggs again. At the end of the first laying cycle the hens will go into a "molt" phase when they stop laying eggs and lose some of their feathers. It's during this time that the hen rests and allows her reproductive system to regenerate. Then she will start laying again for another year.

Many poultry producers don't keep their hens for the second year, preferring instead to start with a fresh batch of chicks and repeat the process.

We start our chicks five months before the farmers markets are set to open in the spring. They will go through the brooder for 2 to 3 weeks, and then on to a Day Range pasture until they start laying at about 5 months of age. Their early eggs will be small "pullet" eggs, but within a few weeks the eggs will increase in size until most of them will be medium to large.

You can buy egg scales, egg cartons, egg graders, and egg washers from poultry supply houses and NASCO (see resource section).

Our feed ration also includes calcium, but we always keep some oyster shell in a special feed trough in the layaway. We also use planer shavings for bedding in the layaway, and throw scratch corn on it daily to encourage the hens to peck and scratch through the bedding. This keeps the bedding fluffed up and the smells down. The hens get any larvae that hatch in the bedding.

Layers, unlike broilers and turkey poults, will always go back inside the layaway at night to roost. We don't have to worry about them getting caught in a cold rain during the night, so unless we have a severe predator problem that the electric fence is not stopping, we don't even bother to close them in the layaway at night. At first light they are back out on the ground pecking and scratching and having a good time. When we bring them the morning feed we also check the layaway for problems, and then put down the roosting bars so they can have access to the nests for laying eggs.

We use a Fortex bucket that is 4-inches deep with a garden hose and float valve for fresh water, and clean it daily. One bucket will serve up to 400 hens. Because we keep hens year round it is necessary to deal with water freezing. The water in the garden hoses and in the Fortex bucket freeze overnight. We have a choice of either shutting off the water flow in the garden hoses and letting them drain every night, hauling water to the birds, or bringing the hens to the barn where we can more easily service them.

We prefer to bring the livestock to the barn for our convenience of taking care of them. For water we use portable water fountains that we empty at night to keep from freezing and refill each day. They hold 5 gallons of water and cost $25 to $35 and are available from poultry supply houses and from Tractor Supply Company. They have a vacuum delivery system that works on sloped ground if you put the outlet on the down hillside. It will fill up and stop flowing before the drinker tray fills completely.

We use plastic nest liners. They are ventilated and clean easily to eliminate nesting sites for predator pests such as mites and fleas. They are available from G&M Sales in Harrisonburg, Virginia (see resource guide).

By keeping the bedding and the nests clean and giving them access to fresh pasture regularly, you will have few, if any, problems with your flock. If you should get a health problem, Gail Damerow's *Chicken Health Handbook* is an excellent resource.

Scratching and pecking and dust bathing are all good things for a chicken or turkey to do, but are a little hard on the pasture. Just go back and fill in the holes with dirt or compost after you've moved the chickens to a new site.

Roosts are made from 2x4 lumber, untreated. Each hen needs 6 to 8 inches of space, and the roosts are 16 inches apart. You can also make them step up in the middle of the room, sort of like a triangle.

The layaway has a plywood floor and sides and a metal roof. The ends are doors that open and are covered with poultry netting for ventilation. In the winter time we staple 6-mil plastic over the doors to protect the hens from cold air, wind, and rain. We use a 4-foot fluorescent light fixture to give them supplemental light in the winter time so they continue to lay enough eggs. The sides are 5-feet high, and the roof peaks up enough that we can walk through the building to gather eggs and tend the chickens.

We move the layaway daily, weekly, or seasonally, depending on where they are and what we want to accomplish in pasture renovation, or for preparing ground for a garden next year.

3. Turkey Flock Management (Meat Production)

The turkey poults will stay in the brooder for up to eight weeks. At a young age they are fragile and tend to die for no apparent reason, although some causes of death are: being too hot or too cold in the brooder, improper feed, too crowded, or any one of a dozen different ailments.

Once they are 8-weeks old they can leave the brooder and go directly on pasture, but they will still need shelter from cold night rains and blazing hot midday sun. Some growers have claimed good results by just providing open roosts on the range, but we have found this is needlessly cruel to the birds, and often results in mortalities. We always give them a covered roost, but in most climate zones, it is not necessary to bring them inside the barn unless you are growing turkeys through the winter. Usually it is not too cold by the time they are harvested for Thanksgiving.

We provide a 20-foot feed trough for each hundred turkeys, and keep feed in front of them at all times. They will avail themselves of the pasture, but for high gains they will need continual access to feed. We provide one bucket drinker for each 200 turkeys. We move the feed troughs and water buckets daily to clean ground. This is a way to steer the birds around inside the paddock for even distribution of the manure and to spread the wear and tear on the pasture.

In the brooder, we keep the turkeys in flocks of 50 to 200, with plenty of room to move around and exercise. After they get their primary wing feathers, at about 4 weeks of age, we clip the tips of the primary feathers on one wing. It's easiest to do this at night when they are bedded down. It's important to clip the wing before they ever learn they can fly, then you won't have a problem with them flying out of their paddock or hopping across the fence.

In the field, we blend the small flocks into one large flock, introduce them to a hot poultry netting at about 8 weeks of age, and give them about 1,600 square feet of pasture per 100 birds. As they get older, you will want to move the pasture more frequently because they put down a heavy coat of manure. It takes 4 lengths of electric poultry netting to provide temporary paddocks for up to 500 turkeys.

In mild weather they will ignore their shelter and just bed down wherever they take a notion in the evening. In the morning when they wake up they almost immediately defecate, leaving a good supply of manure in their wake. You can use that to your advantage by driving them to a particular area where you want to fertilize the pasture, and leave them there overnight.

4. Turkey Breeder Flock (Next Generation)

Why have your own turkey breeder flock? There is certainly no shortage of these big birds at the grocery store. And you can easily order pullets every year for your farm that will give you a good product for your customers. The biggest concern the American Livestock Breeds Conservancy, and we have, is the lack of genetic variety.

100% of the commercially available birds are of a single breed called the "Large White" or the "Broad-breasted White". Because of their large muscles and short legs, these birds can't mate naturally. Therefore, 100% are reproduced by artificial insemination. They are genetically very similar, and the rest of the heritage breeds are statistically insignificant.

Any turkey breed other than the Giant White is rapidly becoming extinct. We feel it is important to preserve the heritage breeds. We also believe it is important to have naturally breeding turkeys to be available to the small producer. It is in the global best interest to preserve the genetic diversity they represent for the future. We are concerned that to the best of our ability, we preserve as much as we can for the next generations.

We have raised heritage breeds of turkeys as well as the high-octane Giant Whites. All have done well on pasture.

Contrary to what we were told – that the Giant Whites would die if left out – we found they were good grazers and thrived on pasture. They did well outside in all kinds of weather, gained weight fast, and dressed out to beautiful table birds. If anything, we found the genetic super-charged Giant Whites have the potential of getting too big too fast. If you let them grow too long they become so big you can't get them into the normal oven, and people don't need such a large bird.

This may have been the reason why the USDA poultry researchers spent nearly ten years developing the Beltsville Small White. Unfortunately, this breed became extinct in less time than it took to create it. We still have other breeds that are important to us though in the White Holland, Bourbon Red, and Royal Palm.

As far as we can tell, the Large White and the Broad Breasted Bronze are just as good at grazing as the smaller varieties. But they can't pasture breed. Enter the Sweet Grass turkeys.

The Sweet Grass History

This history is adapted from an article in the Snood News, published by the American Livestock Breeds Conservancy, Fall 1998, Volume 1, Number 2. Credit goes to Bob Hawes and Carolyn Christman as authors.

In 1993, the Sweet Grass Farm of Big Timber, Montana began a flock of turkeys with the goal to have a commercial operation using the naturally mating broad-breasted bronze. The foundation stock came from two sources: (1) a research flock at Oregon State University which was being dispersed, and (2) birds directly from the Wishard Farm.

In the spring of 1995, 153 poults were hatched that had useful production traits and characteristics. Sweet Grass Farm began investigating the possibility of semi-intensive production.

This group of turkeys exhibited excellent disease resistance, white pin feathers over much of the carcass, and good feed conversion, making it appropriate for production of a chemical-free table bird. The birds were raised on range, where they ate grasshoppers, grass, seeds, and weeds along with grain supplementation.

In 1996, the farm raised about 550 pullets and sold both fresh and smoked turkey to gourmet markets. The meat was well received due to being chemical-free with a price that was several times more than the price for supermarket turkeys.

Customers liked the opportunity to purchase a turkey that was close to a wild turkey in its lifestyle, i.e. naturally mating, foraging, and raised outdoors on pasture.

Ma & Pa,
American Sweet Grass

Customers also supported the program that helps preserve the old domestic form of the breed. At that time, as today, the road block was the lack of a USDA-inspected poultry plant in Montana to process the birds. They had to be transported to California for processing.

Then in 1996, an event happened. Seven light-colored poults were produced by the conventional

Turkey Gothic

Bronze parents. These matured into five hens and two toms. They were bred together and produced only lightly colored, white pin-feathered poults.

Their color pattern is beautiful and looks like a heavily marked Royal Palm, dipped in chocolate. There is reddish coloring in the wings and tails of the males and females. The under color of the feather is white, producing a clean carcass without the melanin pigment in the feather follicles of the Bronze birds. This makes them much faster and easier to dress. Their leg color is also white.

The genetics of the color and its genetic base are still underway. It is likely to have originated with the Oregon State Flock, and it might be related to the pattern identified by Dr. Tom Savage as the Oregon Gray.

We are not sure at this point, but we think it might be a new pattern. Thanks to Phil Sponenberg, Ed Buss, Tom Savage, and Craig Russell for their input on the color and genetic base.

In 1998, more light and bronze poults were produced at Sweet Grass Farms, but the owner has decided not to continue with the turkeys. Through the efforts of Phillip Sponenberg, American Livestock Breeds Conservancy (ALBC) contracted to purchase the remaining 15 adult breeders. They were split into 2 flocks, one sent to Glenn Drowns of Sand Hill Preservation Center in Calamus, Iowa and the other to Andy Lee and Pat Foreman, authors of *Chicken Tractor*. The rest of the story unfolds below…

The Great Sweet Grass Rescue – Late summer, 1998. Enter the authors with funding from the ALBC and Good Earth Publications.

How did we get interested in turkey breeder flocks? We credit Don Bixby and Carolyn Christman of the ALBC. We also credit Chief, a broad-breasted bronze that was one of our first turkeys.

So this rescue begins with a love story...

Once upon a time, when we moved from North Carolina, Chief was the only turkey at Living Earth Farm. And he was lonely. So lonely that he was constantly walking the fence and gobbling ALL day long.

Obviously, he needed company of his own kind. The chickens wanted nothing to do with him. So we got him Morning Star, a Wishard-Bronze hen. When they met, she was aloof and she wanted nothing to do with him. He began a courtship to win her favor, and after a few days, Morning Star gave her consent with a waltz. This was the only time we ever saw a turkey mating dance. The dignified courtship was a ballet, so gentle and graceful that we were awed by it. It was poetry in motion. Chief would strut and embrace Morning Star under his wings and they would circle step in unison clockwise. Then they snuggled and entwined their necks together and continued the dance. This scene was far, far different from the bump and thump that we farmers usually see.

Coincidentally, a few weeks later, we got a call from Don Bixby, Director of ALBC, saying that natural breeding was actually a very rare and endangered thing. He had never heard of such a mating dance. And, by the way, did we have an interest in helping to preserve another rare item – a turkey flock that could breed naturally on pasture?

We replied: "Yup, we'll do that!"

That's how Patricia and I got involved with Sweet Grass turkeys in the first place. We wanted to help save the breed from being butchered in Montana. Honestly, this is a true story. In August of 1998, (bad time to Express Mail mature turkeys as it turned out) ALBC arranged for seven turkeys to be express mailed to us, and seven to Glenn Drowns. One of the two toms sent to us was dead on arrival, and both of the toms sent to Glenn were dead on arrival. This was a big quandary. The birds were so beautiful that we were inspired to help preserve the breed.

Now we had a problem. How can you have a preservation program if you only have one tom? Then Carolyn Christman who was at ALBC at the time, and a good friend of ours, suggested we arrange to ship the birds from the ranch in Montana via a cattle truck or something similar, with enclosed sides.

Alternatively, a plan was for Andy to fly to Montana, rent a truck, and haul the birds home. At that time the flock totaled 148, of which some were definite culls, and some were offshoots of Wishard Bronze. So, that's what we wound up doing.

On Halloween day, Matt Jones and I flew to Billings and rented the biggest, ugliest, fuel hogging-est truck available, and drove on to Sweet Grass Farm in Sweet Grass, Montana, about 60 miles west of Billings.

Our plan was to arrive after dark, sneak the truck up next to the pen where the 148 birds were, and one by one capture them off the roosts and put them in the truck. The ranch manager, Ron Wiggins, and two of his ranch hands met us at the main house, where we proceeded as planned, and all went smoothly. We finished loading just as the most beautiful full moon I'd seen in years came up in the "Big Sky". Matt and I headed for the East Coast, while the ranch hands took their kids trick-or-treating.

To this day, we sincerely believe that if Matt and Andy hadn't shown up that night, those turkeys would have been Thanksgiving dinners for sure. Not because Ron Wiggins wanted to do that, but because he simply had no other choice. You all know how much it costs to feed 148 turkeys, and build shelters for them, and keep them safe from coyotes and bears. Butchering the birds was seemingly his only alternative, since he'd been advertising the flock for a year and had only a few takers of trios.

With additional financial support from ALBC, and a substantial grant from Good Earth Publications, we bought the entire 148 flock. Of these, we culled 15 of the crooked toes (indicating inbreeding). I also sold off the bronze ones because they were too flighty and we couldn't keep them in a pen.

We coined the name Sweet Grass because of three reasons: 1.) The turkeys were from the Sweet Grass Farm in Montana, 2.) They are truly pasture birds – eating grasses, and 3.) Sweet Grass is a ceremonial plant used by Native American Indians and we wanted to honor the heritage of these birds. Besides, Patricia is part Native American and likes those types of things.

Ideally, we think a hybrid cross with a Sweet Grass tom crossed with Giant White hens would produce a good table bird. Genetic work is ongoing at Sand Hill Preservation Center and we hope this breed will become commercially viable one day.

Turkey Breeder Pens

Turkey breeder flocks need special pens during breeding season. Part of the function of the pen is to be able to separate all but one tom from the other males so they won't fight and injure each other instead of paying attention to breeding the females. This also helps us to verify that each tom is actually fertile.

Turkeys will naturally seek the highest place they can to roost, including roof tops. However, high roosts should be discouraged because these heavy birds can bruise or injure themselves when flying down from high places.

We allow more than enough roosting space because we really want our breeder flock to be up off the wet ground, especially in Winter. We also feel this is cleaner for the birds rather than forcing them to settle down in the litter.

Turkeys need elbow room. We rarely see them cuddling up to each other at night on the perches. We allow an ample 2 to 3 feet per bird with the roosts at least 18 inches off the ground. If you use a ladder roost, then begin the first one about 18" inches off the ground and the other perches about 6 to 12 inches above each other. The highest perch should not be more than 6 feet above the ground.

Our barn roosts are all on the same level and are about 2.5 feet apart. This gives the birds enough room to get up between the roosts. We use 2x4 lumber fitted into joist holders so we can remove them for easier cleaning. Joist hangers are galvanized metal fixtures used for holding floor joists against the outside rim boards. They are available at lumber yards and home centers. You could also use 2 to 3 inch diameter poles.

Turkey Nest Boxes

Nests for turkeys are usually on the floor or are raised a few inches. If you use a double decked box, provide a two foot ramp so these big birds can get up, down, in, and out of the nest boxes easily and without hurting their legs.

You should have enough nest boxes so that 1/2 to 1/3 of your hens can lay at any one time. Hens readily take to nest boxes as long as no tall vegetation or other hiding places are handy. Have the nest boxes in place a few weeks before production begins to give the hens a chance to get comfortable with them.

We made our nest boxes out of plywood. The dimensions are: 16-inches wide x 2-feet deep x 2-feet high. We also put in a threshold about 4 to 5 inches high to hold the bedding in. These nest dimensions are larger than the older text books recommend. But, we found that our birds are bigger than they were in the 1950s. I guess the same is true of our people.

We prefer to use wood shavings for nesting materials, but there are a wide variety you can use including hay, straw, leaves, rice hulls, etc. A hen is not picky about nesting materials nor are they picky about nesting sites. We have used saw horses with

Plywood turkey nest boxes. One way we could make this more comfortable for the hens would be to cover more of the entrance to make the nests darker and more private. A 4" board or a cloth across the top would do. A high threshold is essential to keep the bedding in. Not visible are 1/2 inch ventilation holes in the sides and back. We keep the nest boxes in the barn or under shade, never just out in the sun.

plywood on the sides, barrels, and almost any place where a hen can feel secure, have some privacy and a dry, clean nest.

Hens will also nest in tall grass or under shrubs. If you are missing a hen or two, chances are they have established their own nest in the tall grass or some other hidden place. Discourage this for her safety and your egg collection.

Trap Nest Boxes

If you want to know which hen lays which egg, then you will need to use trap nest boxes. Trap nests allow only one hen into a nest at a time and traps them there until you let them out. Collect the egg and mark the hen's number on the small end of the egg. Eggs that get laid outside the nest are called "floor eggs". Floor eggs can be a problem because you can pedigree them only for the tom.

Trap nesting can also be used to determine which hens are laying eggs and which are not. If a hen goes in, gets trapped, and you find an egg when you release her, then mark her so you'll know that she is indeed laying. Hens that don't get marked after several days are not laying and can be culled from the flock.

A trap nest can also help you identify any hen that lays abnormal eggs (oddly sized or shaped). These hens should be removed from the breeding flock. This will help undesirable egg characteristics from being established in the strain. For example, we had one hen that routinely laid thin-shelled, or shell-less eggs that broke easily. This was not a trait we wanted to select for and it was a problem with getting other eggs dirty, so we culled this hen from our breeder flock.

Trap nest boxes require intensive management. The boxes should be checked mid-morning and again in the afternoon or even more often. You will need at least one trap nest for every 2 hens.

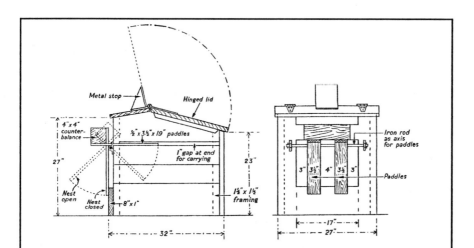

This design is an individual, self-closing turkey trap nest from Ray Feltwel's book *Turkey Farming*. Left shows the cross-section and Right show the front view.

Record Keeping & Tracking Who's Who in Your Flock

Use leg bands to keep track of which hens and toms you want to keep. We recommend metal bands with permanent serial numbers sealed in the metal. Put the numbers on the right leg with the band placed upside down so the number can be read easily when the hen is lifted from the trap nest.

We found the plastic bands are easy to use and read, but come off too easily, especially off the toms when they fight each other. When we use plastic bands we put on two, so that when one falls off we can still uniquely identify the tom.

Genetic Management

There is not a lot of current material available for the small farmer raising turkeys. Ray Feltwell, in his book *Turkey Farming* printed circa 1940s, suggests that: "To allow some measure of improvement, a minimum of at least 100 hens in the breeding stock is desirable". Another valuable out-of-print book that has breeding protocols is Marsden and Martin's *Turkey Management* circa 1950s.

All the texts say that as a general rule of thumb is to allow 1 tom per 8 to 10 hens as a good ratio. We only let 1 tom in with our hens at a time. Two or more toms with the hens spend their time fighting and knocking each other off the hens instead of being romantic. This results in a lower egg fertility and lower hatch rate.

Charles Wishard, founder of the Bronze Wishard strain, started in about 1945, to raise bronze turkeys on pasture. His field breeding program was low-tech and low maintenance. He just kept all the toms with all the hens. He collected and incubated the eggs, and raised the poults to eight weeks, at which point he let them out on pasture to fend for themselves. His range birds were provided with pastures enclosed with woven fence, elementary shelters with roosts and feeding and watering stations which were moved a few yards each week to minimize ground contamination and to rotate the pasture.

The 6th edition of Marsden's *Turkey Management* describes several plans to prevent inbreeding. All these plans involve rotating toms over time from a few days to yearly. Two plans are described below:

In one system, a single tom is rotated through each of several hen pens twice weekly throughout the season. This gives a regular rotation of males which can help fertility where preferential mating is a problem. For example, label your pens A, B, C, D, etc. On each Monday and Thursday, the tom in pen A is shifted to pen B. The one in pen B shifts to pen C, and so on round-robin, With this method, almost every hen is sure to be mated and you will get genetic diversity.

Another common plan for small scale breeders is to separate all the toms from the hens and alternate one tom at a time with the hens. This is done once or twice weekly to maximize using all the best toms for genetic variety. This way all your toms are used, but a single one is never with the hens so long that his genes will dominate.

If you have a superior tom and want to have him sire most of the next generation, then keep that tom with the hens for a longer time. One mating period will last about 4 weeks for a hen.

With the Sweet Grass turkeys, we found the designated batter-up tom pays more attention to his hens if he doesn't have the rest of the toms jeering and challenging him from the next door paddock. With the other males so close, the tom would pay more attention to them strutting along the fence and making hydraulic sounds, and looking as fierce as possible all day long. Meanwhile, the hens were ignored and not mated.

Leaving a vacant paddock between the breeding flock and the rest of the toms not only lets him concentrate more on his job,

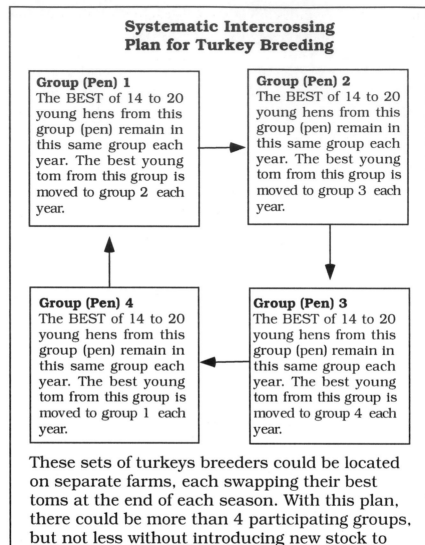

Systematic Intercrossing Plan for Turkey Breeding

Group (Pen) 1
The BEST of 14 to 20 young hens from this group (pen) remain in this same group each year. The best young tom from this group is moved to group 2 each year.

Group (Pen) 2
The BEST of 14 to 20 young hens from this group (pen) remain in this same group each year. The best young tom from this group is moved to group 3 each year.

Group (Pen) 4
The BEST of 14 to 20 young hens from this group (pen) remain in this same group each year. The best young tom from this group is moved to group 1 each year.

Group (Pen) 3
The BEST of 14 to 20 young hens from this group (pen) remain in this same group each year. The best young tom from this group is moved to group 4 each year.

These sets of turkeys breeders could be located on separate farms, each swapping their best toms at the end of each season. With this plan, there could be more than 4 participating groups, but not less without introducing new stock to prevent inbreeding problems.

but was quieter on the farm as well. Ideally, we have the non-breeding toms totally out-of-sight from the hens.

Toms and hens need 100 square feet each in the breeder pen if they can rotate to new ground frequently. If you cannot rotate to fresh pasture, then give them 400 square feet each to keep them from destroying the graze in the paddock. In a larger paddock, just move their shelter, fountain and feeder daily or weekly to spread impact on the lot.

Marsden also describes a system of "systematic intercrossing" which would decrease the danger of close inbreeding. He claims this system would last many years — at least 16 years before needing "new blood". This assumes each mating is done only once. If the breeding stock is selected only from the second season's offspring, then the system will last even longer.

Turkey Egg Production and Natural Mating

Well matured young hens and toms are generally better breeders with higher egg viability than older birds. Age of breeding stock can vary from as early as 20 to 24 weeks and up. Generally toms older than 2 years are less vigorous. Hen egg production decreases each year and is very low by the third production year.

Depending on the lighting and climate, you can expect to get the first turkey eggs around early March to the end of March, depending on your latitude. To minimize carrying costs, the next years breeding stock should be hatched from mid-June to the first of July. This allows the breeding stock to fully develop, yet minimize, the long maintenance period before production starts.

Once mated, a hen will have viable sperm good for about 4 weeks. So if you find a hen or a tom in the wrong pen, the eggs will be of doubtful pedigree for about a month. Generally, a hen will lay about 60 eggs during the season. Of these, we are very happy if we get an 80% to 90% percent fertility and egg viability (hatch rate) with the natural field mating. We have gotten as low as a 50% hatch rate from the eggs we set.

Just as with chickens, peak egg production is reached during a hen's first season. Second year hens lay about 20% fewer eggs, with a similar drop during the 3rd year. Hens older than 4 years old will lay eggs, but the number might not be enough to merit the feed and carrying costs to keep her. However, for the minor heritage breeds whose populations are so small to begin with, and for very valuable outstanding individuals, it might be worth your while to keep older breeding stock.

Selection of Turkey Breeding Stock

We want to stay with the broad-breasted birds because they have more white meat, which customers like. We want to stay with the lighter feathered birds because the pinfeathers on the darker birds are really hard to get out when we process them. The white birds seem to yield a cleaner, more attractive carcass that shows and sells well.

The industry has a lot of quantitative, scientific measurements and tracking methods that they use. But in our small, pasture raised, naturally mating flock, we just stick with these general guidelines. The way we select our breeding stock is rather subjective, but it seems to work for us. Beginning with new breeding stock from 22 to 30 weeks old, and constantly scanning our older birds as well, we look for the following points:

a. Muscle Development. Fullness of Muscle – Breasts, Leg, and Back

You can determine this by handling. The breast should be smooth and broad with the width carried well out to the rear of the breast bone. The breast will taper slightly towards the rear of the keel. This is normal and necessary for movement and correct leg location. The breast muscles can be too narrow at the rear, or too broad and square. The breast should be free from knobs, calluses or breast blisters.

The thigh and leg (drumstick) should be plump and well muscled towards the hocks. The legs need to be big enough to carry the body with good balance.

b. Skeletal Development

The neck should be reasonable short, and the legs reasonably

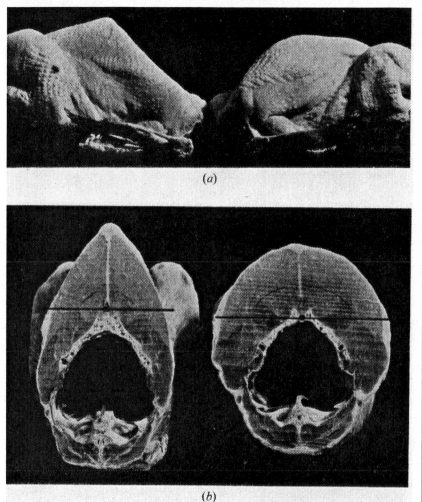

Commercial Broad-Breast vs. the Narrow Keel-Breast Meat of a Wild Turkey

(a)

(b)

Broad-breasted birds carry a greater amount of edible meat than the narrow and high breasted turkeys. (a) Side view of high breasted compared with a broad breasted bird. (b) Cross section showing the great difference in edible meat. Most heritage and wild turkeys are not broad breasted. Photos adapted from Ray Feltwell's Turkey Farming.

long, but not leggy, as with the wild turkeys. The body needs a good width at the ribs and over the hips. The back should be relatively broad with a well-sprung rib cage.

When held upside down, the keel (breast bone) should be straight and parallel with the back extending backward toward, or between the legs. There should not be a notable gap between the legs and the rear point of the keel. This shows the balance of the entire body.

c. Carriage and Motion

You can also observe the bird's walk. Look for balance and ease of motion. Good balance includes an upright carriage with the back sloping about 35 degrees above the horizontal. An undesirable carriage is when the body is tipped forward and the breast is carried very low. This puts too much strain on the hip joints and lower back.

d. Legs and Feet

The broad-breasted birds carry more weight, putting an extra strain on the legs. The result is a tendency toward various kinds of leg problems. Select for strong, straight legs with well-formed hocks and muscled thighs.

Look for any other leg abnormalities, including in the toes. Toes curled abnormally one way or the other are a symptom of inbreeding. Unless you think the bird broke it's toe, exclude it from your breeding program.

e. Movement

Get behind or in front of the bird and observe how it walks or runs. Look for and cull all birds that:

- Throw one or both legs out to the side (paddle or toe out)
- Appear bowlegged
- Limp noticeably (if so, look for a wound or bruise)
- Show any kind of crookedness to the gait or foot prints

Nearly all broad-breasted turkeys will waddle a little with

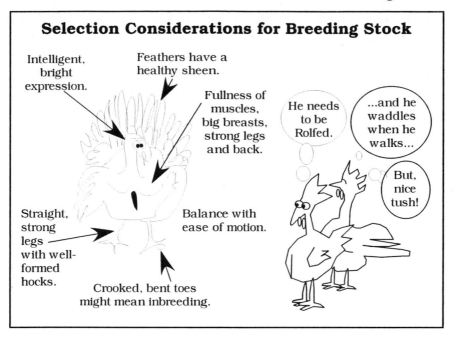

a sidewise movement. Look for their feet to track almost straight ahead of the last foot pint and not thrown noticeably outwardly.

f. General Condition

Finally, assess if the bird has the general "bloom" and a healthy look to it. Is the skin under the breast feathers pink and creamy? Are the feathers the color pattern you are selecting for? Are there any other defects, deformities, signs of disease, or weakness?

Winter Care of Turkey Breeders

For our convenience, we bring all our breeders into our barn for the winter and we keep all the toms and hens together. This way they are easy to service and monitor. Also, we sleep well knowing they have really good shelter with protection from wind, snow, and rain. We have a 30x30-foot area in the barn with plywood walls with the east end open and secured with chicken wire. We have deep wood shavings bedding and roosts about 3 feet off the ground. During the day, we open the door and let the birds have access to a 2-acre fenced paddock.

It is important to restrict the diet of the turkeys so that they don't get too fat and out of laying condition. On average, a turkey will need only about 1/2 pound of ration per day. For a flock of 30 birds that's only 15 pounds of feed per day. It might seem that you are starving them because of the way they rush up and gobble down the feed. But it is important to have them fit and in breeding condition when the season begins around March. In colder weather we give them additional corn along with their ration to help them generate body heat and stay warm. Some of the older texts also say to feed them meat scraps.

Use your common sense. If individuals in your flock look too thin or unhealthy, then make sure they have access to their fair share of feed. Sometimes we have had to separate some birds to give them a chance to gain weight and get in better condition. Let your birds tell you how much feed is enough, which you can observe by their condition, body weight, and muscle tone.

5. Broiler Breeder Flock (Next Generation)

We manage our broiler breeder flock very similar to our egg layer flock, including the same feed ration. There is one major management difference between laying hens and broiler breeders. This feature can't be emphasized enough; carefully control the amount of feed you give these broiler breeder birds. They are genetically programmed to gain weight and gain it fast with very good feed conversion. We give layers free access to all the feed they want.

But, If you overfeed broiler breeders, they will get fat and become egg bound – a problem that we described earlier. Otherwise, the housing and routine is very similar to layer flock management, except most of these eggs go to the incubators.

The reasons for having your own broiler breeder flock and hatching chicks on your farm include:

1. "Pasture Peeps" can provide a good supplemental or primary farm enterprise.

2. Some shipping companies, for example FedEx and UPS, will not carry chicks.

3. You can schedule hatches and production rates without worrying about weather and temperature conditions that dictate commercial transport.

4. "Pasture Peeps" might be hardier than ordinary broiler chicks.

5. Bio-security advantage of not importing potential disease carrier birds on to your farm.

6. Chicks are not stressed in long-distance shipping.

7. Development of strains for regional conditions.

In the next chapter we discuss in detail how to set up and manage an on-farm hatchery.

Especially during the Holidays, the Sweet Grass show how much smarter thay are than the average turkey.

Chapter 9: An On-Farm Hatchery

Building an on-farm hatchery for broiler chicks makes a lot of sense. It can be fun, affordable, and practical. High quality incubators are available for reasonable prices. Growing a breeder flock is no more difficult than taking care of a layer flock. Our broiler chick cost from our own hatchery is about 30-cents less per chick, which is about a third of the cost of buying chicks from a distant hatchery.

However, having an on-farm hatchery for layer hens or for Thanksgiving turkeys might not be such a good idea. If you are hatching the layer breeds, you will need the greatest percentage of chicks in the spring. To hatch layer chicks, means you'll have to throw away 1/2 the birds hatched that are males, as they have no commercial value. Sexing chicks is easier if you use hybrid crosses that are sex linked. Sex linked implies that male chicks will be a different color from the female chicks.

Turkey poults are a particular problem because they are all in demand in July, but not so much the rest of the year. You can buy eggs, rather than have your own breeder flock, or raise rare and minor breed turkeys and sell the poults to fanciers. Broad breasted turkeys hatched during mid-July will be just the right size for Thanksgiving, 16 weeks later.

"If you do not have problems,
you can miss opportunities for growth",
Chairman Mao Chicken-Tung's Little Egg Book

Earlier, we talked about breeder flock management for additional information on the care of chicken and turkey breeder flocks. This chapter focuses on a small-farm hatchery.

Here are the reasons why you will want to consider building your own on-farm broiler hatchery:

- Additional income
- Local or regional supply
- Easier on-farm production scheduling and season extensions
- Compatible with other farm chores
- Healthier chicks and biological security
- Genetic diversity and few reliable hatcheries to select from
- Many commercial hatcheries don't have a breeder flock they buy eggs as needed on the open market.

Now lets examine these reasons more closely.

Additional Income from Incubating

Probably the greatest reason to hatch and raise your own broiler chicks in an on-farm hatchery is that it can be very profitable. There are two income streams with this operation. The first is that you will save money on your own chicks to grow out at your farm. Second, you will be able to sell chicks to other growers in your region. We estimate it costs about 25-cents per chick to have our own hatchery. We can sell the chicks for 85- to 95-cents each, and the buyer pays the postage, or comes and picks up the birds at our farm.

Here are some approximate costs to maintain a broiler breeder flock.

Breeder Female chicks	$4 each
Breeder Male chicks	$3 each

(You will need 1 rooster for each 10 hens)

Feed purchased in bulk cost 12-cents per pound with each hen eating about 110 pounds of feed during her 64 week life span. That results in a total of about $13.20.

Labor will cost about $5 per hen per year.

Breeder flock shelter and hatchery equipment will cost about $10,000, amortized over 10 years, this comes out to about $5 per hen per year.

A typical on-farm hatchery might consist of 200 broiler breeder hens and 20 or 30 broiler roosters. We purchase these as day old chicks from breeder hatcheries. The breeder chicks cost a lot of money, but it's the only way you can ensure their offspring will grow out to market size and condition in a reasonable length of time on a predictable amount of feed.

We conducted informal on-farm experiments to determine if it is absolutely necessary to go back to the breeder hatchery each year for new breeder stock. For two years we held back some of the regular white Cornish Cross broiler chicks we had purchased from commercial hatcheries and grew them out to be breeders.

These were the first generation (F1 hybrid) meat birds. Because we were inbreeding them we expected some of the off spring not to be true body type and color. As it turned out, about 10% of the offspring were gray, and 1 or 2 from each batch were layer types with reddish-orange colorings. Not surprisingly, from these hybrid F1 parents we did not get the same uniform body type in the offspring as we had from their pure-strain original parents that we had bought from the breeder hatchery.

This is not to say that you can't hatch broiler chicks from breeders that you save from the previous flock. For your own customers, it will be fairly easy to handle the diversity of body confirmation and color. But, it probably isn't something you want to do if you plan to sell broiler chicks to commercial growers.

They will expect uniform sizes and consistent color and body types.

Here are some approximate figures for a hatchery of 200 broiler breeder hens and 20 roosters:

Expenses:

Breeder chicks 200 x $4 per chick =	$800
Males 20 x $3 per chick =	$60
Shelter and hatchery =	$1,000
Labor =	$1,000
Feed (110# x 12-cents/hen) =	$2,900
Shipping boxes 300 @ $3/box =	$900
Total cost	$6,660

Income:

200 eggs per hen x 200 hens = 40,000 eggs

- 10% eggs not set =   -4,000 eggs

Set 36,000 eggs to expect a 75% hatch = 27,000 chicks

Each chick is valued at .85 each = $22,950

Gross margin = $16,290 (not including farm overhead and management costs).

You might be able to sell the breeder hens as "stew hens" at the end of their laying cycles. Each stew hen will be worth about $5-$7, unfortunately, there isn't a very large market for stew hens, although some restaurant chefs will value these older birds for coq au vin, a very tasty French entree that requires a mature chicken, stewed in wine with onions, mushrooms, and bacon.

Local or Regional Chick Supply

Every so often, we get an alarm letter about airlines and delivery services discontinuing the delivery of live animals. Currently, there are only three airlines that will carry day-old poultry, and if those three change their policy, it would force most of the commercial hatcheries out of business.

This is a quote from a letter sent by Murray McMurray to all their customers: *"None of us who ship birds for a living can continue to exist on a year-to-year basis never knowing when and if the airlines will "pull-the-plug" on us. Those of you who depend on people who sell birds through the mail, need to assess where your business will be if day-old chicks can no longer be shipped by mail. Will you still be in business? How severely reduced will your annual sales be? Can you really afford not to get involved in this effort?"*
— Murray J. McMurray,
owner of Murray McMurray Hatcheries and
Chair, Bird Shippers of America.

Currently, Federal Express and United Parcel Service will not carry live poultry. Each year fewer and fewer airlines are agreeing to carry live poultry for the US Postal Service. Fully half of the commercial air line companies no longer carry live poultry.

These commercial hatcheries ship millions of day-old poultry and gamebirds each season. For them to have to rely on a regional market, within truck range, would mean a severe reduction in their sales capacity, probably putting many of them out of business.

In the future it may become important for each region, or even areas within a region, to have an independent hatchery. Within the regional shipping area, USPS can transport live birds via ground mail, thus not having to rely on airlines at all.

Chicks and poults suffer needlessly in air shipments, from change in air pressure and temperature, long waits on loading docks and between flights, and lack of quality air to breathe. Let's face it, nobody has any control whatsoever over the US mail service. Just because the USPS advertises Express Mail in 24 hours, doesn't mean they always do it.

We have shipped birds from our farm with the utmost assurance from our local post office that the birds will be delivered no later than 3 p.m. of the following day. Then, 24 hours later when the birds haven't arrived, we call our post office and they can't locate the package. We call the "tracking number" for the US Post Office and a pleasant recorded voice confidently tells us our "package is on its way to its delivery destination". Meanwhile we know that the live poultry we shipped is slowly dying from dehydration, overheating, or chilling on a loading dock somewhere.

How the Post Office can lose a box that measures 2x2x3-feet and is marked LIVE BIRDS and EXPRESS MAIL, is beyond us, but it happens. Within the close circle of fellow poultry growers here in Virginia, each of us has a story to tell about lost shipments, birds dead on arrival, and midnight runs to rescue stranded poultry from a post office loading dock halfway across the state.

Easier On-Farm Production, Scheduling, and Extending Seasons

When you are ordering chicks and poults from far away hatcheries, you have to operate within the dates that they set eggs to hatch. If you have your own hatchery, you can set eggs daily or weekly to suit your needs. You can time your hatches to coincide with your production schedule. If you need fewer chicks on any given week, just set fewer eggs. If you need chicks earlier or later in the season, just time your breeder flock accordingly.

Early or late production scheduling can be a problem if the hatchery you are dealing with is in a different climate zone than you are. For example, a hatchery in Minnesota will not be able to supply chicks in March to a farm in Alabama. Likewise, a Florida hatchery will not be able to ship eggs after about June, when Florida air temperatures climb above 85 degrees F and the airlines refuse the shipment.

In these cases it becomes very clear that a reasonable solution is for each region of the country to have a reliable hatchery. Failing that, it becomes necessary for each farm to have their own on-farm hatchery. It also makes sense for a group of farms to invest jointly to create a regional hatchery that can supply their needs. We'll discuss this cooperative hatchery concept more in the marketing chapter.

Compatible with Other Farm Chores

The routine of taking care of a breeder flock and a small-scale hatchery is relatively simple, and very compatible with other farm chores. If you are already feeding laying hens or broilers, it's an easy matter to take another bucket of feed along to feed the breeder flock.

Gathering the eggs and sorting them for hatching each day only takes a few minutes of time. Eggs that are cracked or otherwise unusable, or that are infertile, make good pig feed.

Every seven days you will want to candle the eggs in the incubator. Remove those that are infertile, and feed them to the pigs. On the 18th to 20th day, move the eggs from the incubator to the hatcher. Three days later the chicks will pip. Then you can move the new chicks from the hatcher to the brooder. Running your on-farm hatchery will be simple, fun, very cost effective, and very easy to learn. We discuss incubation in more detail later in this chapter.

Healthier Poultry and Bio-security

When you have the parent birds on good feed and good pasture, they will naturally lay viable eggs that hatch into healthy chicks. We have found over the years that chicks and poults we hatch from our on-farm breeder flocks are healthier, more aggressive foragers, and less prone to leg and heart maladies than commercial birds we order by mail. This is probably due the super-nutritious organic diet we feed our breeder flocks, and the chicks having less stress in their first few days because they are not shipped.

Additionally, if we raise all of our own chicks we can eliminate many concerns about importing poultry diseases onto our farm. This isn't to say that you will never experience a disease problem, but your exposure will be limited by the number of vectors entering your farm. You will, of course, want to make sure that any breeder flock you buy has been tested or vacci-nated for the more common diseases.

As part of your bio-security program, you must always be sure that visitors to your farm have not been at another poultry farm that day or that they are not wearing the same clothing. In extreme cases it may even be necessary to rinse off a feed delivery truck tires and undercarriage if that truck has been on another poultry farm that day. Some disease can infect an entire flock overnight, so it isn't worth taking chances, even if you have to go to seemingly extreme measures to protect your farm.

You must always be responsible, too, about spreading diseases from your flock to others. If your hens are infected, it's not fair to continue to sell the chicks from those hens. Spreading a disease unknowingly is one thing, but doing it even after you know the birds are sick, is totally irresponsible.

Be very wary of accepting free poultry from other sources. Last year we adopted an orphan hen, a rare Egyptian Fayoumis, because she was so attractive and needed a home. Within a month our entire flock was infected with poultry sinusitis which we suspect might have come from that one hen, but we will never know for sure.

Genetic Diversity

Over the years the Cornish Cross broiler strains have acquired the reputation of being hard to grow due to leg and heart problems. We have found this is true only if the broiler chicks are not given what they need in order to survive and thrive.

Within broiler breeds there are currently less than a dozen strains that make up the entire population of Cornish Cross broiler chicks. The more commonly known broiler strains are Peterson, Ross, Arbor Acres, Hubbard, and Cobb. Some hatcheries will have more than one strain, and some breeders will cross between strains to get more hybrid vigor.

Some breeders are attempting to introduce more vigor and survivability by introducing genetics from outside the Cornish Cross regime. This is meeting with only limited success. In all probability, the easiest way to ensure survivability is to take better care of the flock. According to poultry breeders, it takes

Incubation Times for Various Poultry Hatch times can vary depending upon temperature					
Fowl	days	Fowl	days	Fowl	days
Bobwhite	23	Eagle	42-49	Peafowl	28-30
Button	16	Emu	58-61	Pheasants	23-26
Canary	18	Finch	14	Pigeon	17
Chicken	21	Geese	28-31	Quail	23-24
Chukar	23	Guinea	28	Rhea	35-40
Cockatiel	21	Mynah	14	Swan	37
Coturnix	17	Ostrich	40-42	Turkey	28
Cranes	30	Parakeet	18	Vulture	28-29
Dove	14	Parrots	28		
Duck	28	Partridge	22-23		

up to 15 years and $150,000 to develop a new strain. Amateur breeders are continually trying, however, and who knows, maybe one day one of them will be successful.

Incubating Eggs & Incubator Basics

We like incubating eggs. Each egg that hatches is a little miracle. The chicks are so cute. Every farm visitor gets excited to see chicks hatching out and hold the babies in their hands. On the business side, it is one of the quickest turnaround and most profitably rewarding activities on our farm.

Incubating eggs is not hard, but there is a lot to know and monitor in the process. If you are going to try incubation, it is well worth your time to study the process. Many little lives in those eggs depend on you knowing what you are doing.

The incubator factors that affect the success of your hatch are temperature, relative humidity, ventilation, and turning the eggs. The first two-thirds of the incubation period is the most critical period for temperature and humidity. Let's look at each of these variables in detail.

Temperature. Chicken and turkey eggs need a relatively constant temperature of 99.5 Farenheit to develop. Temperature is probably the most important factor. Too hot or too cold will affect chick development.

Humidity & Wet Bulb Thermometers. Having the correct humidity is really essential to a good hatch. We use an automatic gravity feed system sold by GQF (Georgia Quail Farm). This consists of a five gallon bucket with a plunger that fits in the water tray. We then use humidity pads to increase the humidity - especially during the hatch.

A wet bulb thermometer is nothing more than a regular thermometer with a wet wick attached. As the water evaporates from the wick the thermometer is cooled. If your incubator air

This is one of our tabletop incubators. The automatic, five gallon waterer for humidity is labor saving. The table needs to have enough space to set a tray down and candle eggs. The clear door gives you a tremendous amount of information about the hatch at a glance, without opening the door and loosing precious heat and humidity.

is on the dry side, a lot of water will evaporate from the wick causing a drop in temperature. Just the opposite for humid air – not as much water evaporates from the wick, and there is less of a drop of temperature from the ambient air.

Soiled wicks become a problem with your wet bulb reading because as the water evaporates the residues and deposits left behind will give you a false reading. As these residues build up on the wick, there is less surface area for the water to evaporate from. This causes the reading to be higher than it actually is and the incubator air will be drier than the reading on the bulb indicates.

Some ways to avoid readings from soiled wicks are:

- Wash the wick about every 2 weeks. Once it becomes crusty, slimy, or discolored it will not give you an accurate reading. Just get a new wick; they are not expensive.
- Use distilled water for the wet bulb wick reservoir.

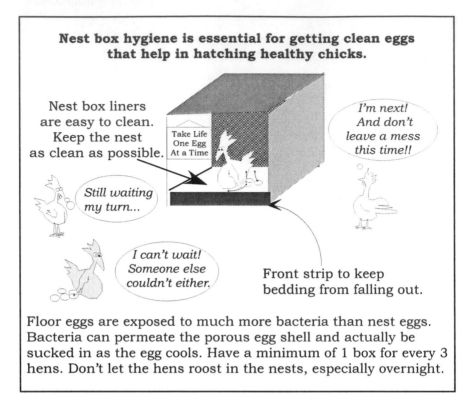

Nest box hygiene is essential for getting clean eggs that help in hatching healthy chicks.

Nest box liners are easy to clean. Keep the nest as clean as possible.

Take Life One Egg At a Time

I'm next! And don't leave a mess this time!!

Still waiting my turn...

I can't wait! Someone else couldn't either.

Front strip to keep bedding from falling out.

Floor eggs are exposed to much more bacteria than nest eggs. Bacteria can permeate the porous egg shell and actually be sucked in as the egg cools. Have a minimum of 1 box for every 3 hens. Don't let the hens roost in the nests, especially overnight.

- Keep the water reservoir clean – it is a place bacteria love to grow.

- Keep the distance from the thermometer to the water not more than 1 inch. If the water is too far from the thermometer then most of the water will evaporate before it gets to the wet bulb and you will get a false high reading.

Ventilation. Air movement and exchange is important because the eggs give off carbon dioxide which can build up. Many incubators have a fan in them that circles the air around. Still air incubators do not have fans but the air moves slowly because of convection.

The main challenge with these still air incubators is that there can be a large temperature difference between the top and the bottom of the unit—as much as 5° F. This is not an insurmountable problem if the temperature is set just right. Be sure to follow the manufacturer's guidelines in setting the temperature, because if you measure the temperature elsewhere your reading might seem too high or too low, but on the average it will be okay for your hatch.

Turning Eggs. Must you turn eggs? It is true that if you set chicken eggs, a small number of them will hatch without being turned at all. Research by Rob Harvey at Birdworld found that each species of birds has an optimum egg turning technique. He even found that some species' eggs won't hatch unless turned and laid on their side.

The bottom line is this: you can hatch chicken eggs without turning them, but you won't get as many chicks. Your hatch rate will be better if you turn the eggs routinely at least twice a day if you are doing it manually, or every 2 hours with automatic turners.

The Incubation Room

It is important that the room where you keep your incubator be able to hold a somewhat constant temperature. Uninsulated buildings, like garden sheds, can have a wide temperature swing and are not usually good for incubators.

When we first began incubation we put our two GQF Sportsman incubators in our downstairs bathroom. These incubators can incubate up to 270 chicken eggs at a time. We put some plywood over the bathtub to make a low shelf which gave enough room for the two cabinets. With the door closed and window slightly open for fresh air, this room has the most even temperature of any in our house.

Not only that, visitors were thrilled to use the facilities and come hurriedly back, wide-eyed, to tell us we have babies. We were amused at how many times folks had to use the bathroom. We truly had a "loo with a view" and no need for reading material. Kids loved it!

Having this close proximity also gave us intensive exposure to observe the hatch and gain experience with monitoring the incubation process. In hindsight, it is one of the best ways we could have learned the art of incubation through observation and easy, multiple-daily contact with the incubators.

That first year with a smaller incubator gave us the confidence and experience to go with a larger incubation and brooding operation. We set out to be self-sufficient in chick production for our farm needs, with a few extras to sell.

We moved the incubators to an insulated room in the barn. We now had three GQF incubators and one hatcher. Each week we would batch set an entire incubator full with about 270 eggs. From this, we hatched about 200 chicks every week. You can buy three incubators and a hatcher for about $1,200.

We also bought a used, 5 layer electric battery brooder. This would hold 500 chicks and made our brooding easier and our losses almost nil. There is a photo of the battery brooder in the next chapter.

Collecting Eggs to Hatch

Getting good eggs to hatch begins with getting and maintaining healthy parents and all that is involved with breeder flock management that is covered in Chapter 8 Flock Management. Here we focus on proper care of the egg the instant it is laid. Collecting eggs two or more times a day will help assure you:

- Get clean eggs to start with, which minimizes cleaning.
- Keep eggs from getting damaged and dirty as the hens get in and out of the nest box.
- Keep the hens from becoming broody.
- Help prevent eggs from being laid on the floor.
- Prevent too many eggs laid in one nest, thereby getting knocked together and cracked.

Normally, we gather eggs as part of the evening chores around 4pm. Most hens will lay their eggs between mid-morning and late afternoon. On days that are extremely hot or cold, we collect eggs twice a day around noon and early evening. This keeps them from freezing overnight or overheating during the day. Eggs exposed to extremely hot or cold temperatures have reduced hatchability.

Good eggs begin with nest hygiene. A dirty nest is like a filthy nursery. Literally billions of bacteria are waiting to attach to a wet egg as it is laid.

Once laid the egg begins to cool, thermodynamics sucks air and any bacteria present through the shell into an almost perfect growth media. The bacteria may not immediately kill the embryo. But like the common cold, bacteria will incubate with the embryo and show up as a yolk sac infection or some other ailment.

Most of us assume that if a chick dies within a week of hatching it was caused by something external after the hatch. In fact, the mortality might have been caused by a problem of contamination three weeks earlier at the time of egg laying.

However you do it, manage your flock and egg collection so that the eggs are as clean as possible and so the eggs do not need cleaning. Eggs naturally have a protective water soluble coating that helps protect them from bacterial infiltration through the porous shell. For lightly soiled eggs use a fine sand paper to lightly scrape and "dry clean" them. For valuable, heavily soiled eggs if you must use water, use cool to warm water. Hotter water can force bacteria through the porous shell and ruin the egg.

On a larger scale, commercial hatcheries routinely clean their eggs using a special bath that replaces the protective covering while still cleaning the egg. We have experimented with both and find that for our small-farm purposes, getting clean, non-

This is one of three metal racks that fit in the incubator. On the right is a plastic egg tray and on the left is the paper egg holder. This also shows the simple candler.

floor eggs initially is good enough and takes that extra bath step out of our production. If we for some reason don't pick eggs daily or get dirty eggs, then we will wash them in the commercial bath solution.

Holding Eggs Before Incubation

Eggs must be fresh and fertile. We try to not hold eggs longer than 7 days before setting. After even a few days, an egg's hatchability declines about 1% per day. By day 14, only about 50% of the eggs will hatch, and by day 21, viability will be almost zero. If eggs are stored without turning, the yolks tend to float to the top which can be fatal. Rob Harvey, in *Practical Incubation*, details how he turns his stored eggs an odd number of times each day. This way they lie on a different side each night.

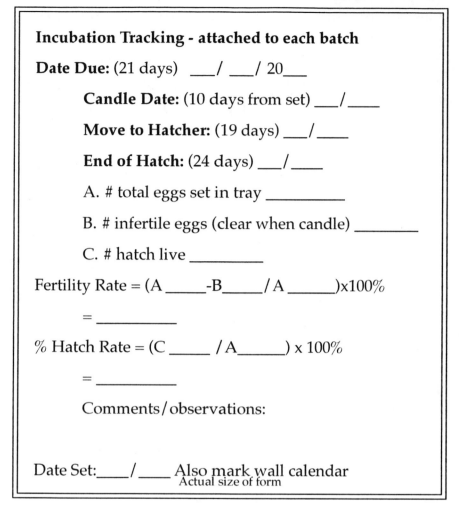

Incubation Tracking - attached to each batch

Date Due: (21 days) ___/___/ 20___

    Candle Date: (10 days from set) ___/____

    Move to Hatcher: (19 days) ___/____

    End of Hatch: (24 days) ___/____

    A. # total eggs set in tray _____

    B. # infertile eggs (clear when candle) _____

    C. # hatch live _____

Fertility Rate = (A _____ -B_____/ A _____)x100%

    = _____

% Hatch Rate = (C _____ / A_____) x 100%

    = _____

    Comments/observations:

Date Set:___/____ Also mark wall calendar
Actual size of form

Ideal storage temperature for eggs awaiting incubation is 55 degrees Fahrenheit. A safe temperature range is 40 to 60F (13C). At room temperature (68F) the egg will slowly begin developing. If left too long, this will kill the embryo or weaken it such that it dies before hatch. The temperature for holding eggs needs to be constantly below embryo physiological zero – which is the temperature at which development does not happen. Any temperature above this will cause some development to begin.

Because of more consistent temperatures and humidity, basements or refrigerators set at 55F degrees can be good places to hold eggs.

Temperature shifts affect egg viability. If you collect eggs that are cold, let them slowly warm and stabilize at room temperature before putting in the incubator.

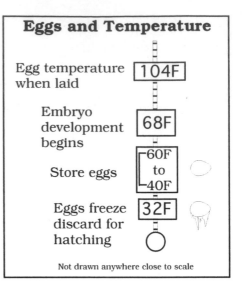

Eggs and Temperature

Egg temperature when laid — 104F

Embryo development begins — 68F

Store eggs — 60F to 40F

Eggs freeze discard for hatching — 32F

Not drawn anywhere close to scale

Keep the relative humidity between 70 to 90%. Humidity is critical; the less moisture loss from the egg before it is set for hatching the higher its chances to hatch. Gail Damerow states in *Chicken Health Handbook* that you can increase the storage time to as long as 3 weeks by wrapping each egg or each flat in plastic — a good use for those extra plastic grocery bags. We rarely hold eggs over 10 days. We keep them stored in a dedicated refrigerator set at 55 degrees. We also sort and store eggs in the flats and in plastic grocery bags. When ready we can transfer them to the incubator with out further handling. We always let the refrigerated eggs warm to room temperature before putting them in the incubator. This way the temperature shift is gradual and easier on the embryo.

Batch & Hatch – Setting Eggs

We find it is easier to set a batch of eggs with at least 1 trayful (90 eggs for our incubators) or more. Some incubators hold 5,000 eggs or even more, so we are really small-scale in our operation.

Put eggs in the holders with the pointed end down. This is very important because it orientates the air sac and yolk for embryo development. We discard small or mis-shapen eggs along with those with cracked or thin shells.

We "batch set" our eggs so that many chicks or turkey pullets hatch out about the same time and can be reared together. This way the babies grow at the same rate. Batching eggs helps us manage our time and focus our attention more completely on each hatch.

To set eggs in our GQF Sportsman incubators we prefer to use the plastic trays. A set of 6 trays is enough for all three racks of our incubators and costs about $30. These plastic egg holders have the advantage of allowing much better air circulation than the paper holders. They are also easier to clean and sanitize. We calculate they pay for themselves in a short time and require less time when setting eggs.

Before we had the plastic egg racks, we used the cardboard egg flats that hold 30 eggs each. We filled in any spare space by cutting up regular cardboard egg cartons. We trim the edges of the cartons so that we can fit in more eggs, and yet still have padding between each egg and along the sides so the eggs wouldn't slide or shift when the rack is turned. We put holes in the bottom of the paper cartons to allow better air flow. You can get egg flats for free from almost any deli or restaurant. If you are worried about bio-security then you can buy new ones from poultry supply companies.

We can set up to 90 chicken eggs per tray and 270 eggs per incubator. Turkey eggs are bigger than chicken eggs, and jumbo egg cartons hold turkey eggs fairly well. If we combine turkey and chicken eggs, then we set every other one with a turkey egg and one of the smaller chicken eggs in between.

We put some packing along the sides of the incubator trays so they won't slide when the tray turns, potentially causing the eggs to impact each other and crack.

Never set cracked eggs. Even hairline cracks will let bacteria invade an egg. Dirty and cracked eggs have a higher chance of absorbing bacteria and becoming rotten fast. They will explode in your incubator. This not only smells literally like Hell — it contaminates other eggs. If you see an egg that is "oozing" and has that sulfur, rotten egg smell, then don't touch it! We very carefully remove the eggs around the egg bomb and dispose of (bury) the entire holder and offending egg. Your nose is the best indicator for bad eggs. If you smell something spoiling find which egg it is and dispose of it immediately. Don't wait! Candling at days 10 and 19 will identify eggs going bad and help keep this problem at a minimum.

Hatching Records

We tried a lot of complicated forms to track our hatching statistics and finally found the easiest way is to simply use a large calendar on the wall by the incubators along with a data sheet we attach to each tray or incubator (see form on page 192). We then enter this batch data on a spreadsheet and let our computers do the calculations.

To track each hatch, we use the label/form that we attach to each batch. We track life-cycle data on each batch of chickens. We combine data from the hatches with their growing statistics (feed consumed and morbidity/mortality), along with processing information (weight and numbers) to track the performance of each batch.

Gathering, analyzing, and using this information takes more time and diligence, but not that much more time. This gives us extremely valuable information to fine-tune our management, identify problems, point out trends, and finally, tells us if we have made any profit at the end of the year for all that hard work we put in.

As an example of how this information was useful, we suspected that something was amiss with our custom-mixed feed ration. We were not getting the expected feed conversions, the viability dropped suddenly in the eggs we were setting, and the chicks that did hatch didn't have the usual pasture peep bloom and vigor. As it turned out, our feed mill had, without telling us, substituted some of our ingredients for cheaper replacements and used old corn that we suspect had traces

Atilla, while usually attentive, forgot to candle her eggs and had one explode in her incubator. A hard lesson indeed!

of aflatoxins. Our information systems raised the red flag that something was not right. This gave us an edge to fix the problem before it did too much damage.

Getting back to information systems, for incubating, we track three statistics, along with constant thoughtful observations:

1. How many and what kind of eggs set.

2. How many of the eggs are infertile when we candle. We usually candle chicken eggs on days 10 and 19, and turkey eggs on days 10 and 26. The second candling is 2 days before hatch, as we put the eggs in the hatcher.

3. How many live chicks we get from each hatch.

4. Any abnormalities or problems we observe (see below on hatch problems).

From these numbers we calculate three indicators:

Hatch rate — which lets us know how our overall system is doing.

Fertility rate — which gives us insights about how our breeder flocks are performing.

Viability of fertile eggs — which gives us feedback on our egg and incubation management.

The myth is true! There is life after egg!

Laying eggs on their side during the hatch helps the chick pip through the shell and exit. It also allows fluids to drain

Even turkeys experience paradigm shifts!

Let's use actual numbers from one of our batches. For this batch we had:

- 270 eggs = total set in batch.
- 33 eggs were infertile and discarded. That left:
- 237 fertile eggs. Of those, we got
- 197 healthy chicks that hatched on time.

Now let's do the calculations.

Hatch Rate = (#chicks hatched live/total #of eggs set) x 100%

For our sample batch: hatch rate = (197 live chicks hatched / 270 total number of eggs set) x 100%

= 72% hatch rate - acceptable, but we prefer higher

We want as high a hatch rate as possible. 72% is a mediocre rate with such basic equipment as the GQF Sportsman. More expensive equipment with better environmental control can achieve hatch rates as high as 98% of fertile eggs set.

We have gotten as high as 85%, and as low as 49%. A low hatch rate is a red flag that tells us we'd better look more closely at our entire operation.

Fertility Rate = (total eggs set − number eggs infertile)/total eggs set) x 100%

For our example: ((270 total eggs − 33 infertile eggs) / 270 total eggs) x 100%

> = (237 fertile eggs/270 total eggs set) x 100%

> = 87% fertility rate - this is very good.

If our fertility rate is low, we look to our breeder flock to see what might be causing the problem. Do we have too few roosters? Feed ration not optimal? With a low fertility, there are a lot of possibilities of things we could adjust with the parents (see breeder flock management).

Our handling of the eggs could also be a factor. Did an egg get badly chilled or stored at an improper temperature? Did the parent start to sit on the egg a day before we collected it? In these cases, the egg was fertile, but the germ might have died and was too small to see when we candled.

Viability of Fertile Eggs = Total live chicks / (total eggs set - number infertile eggs)

> = (197 chicks/(270 #eggs set − 33 #infertile eggs)) x100

> = 197/237 x 100

> = 83% fertile egg viability - this is OK , but not great.

If our viability rate for fertile eggs is low, we look first at how we collect, store, and incubate the eggs. Are we setting eggs that we held too long? Are we collecting eggs routinely but letting them get too hot or too cold? Are the eggs dirty (getting infected) or not cleaned properly? Was the temperature and humidity in the incubators constant and correct? If everything is according to our SOP (standard operations procedures) then we look at the breeder flock and especially at their nutrition. Sometimes extreme heat in the summer will drop the fertility rate.

Thoughtful Observations. Just because a chick hatched doesn't mean everything was correct. You can get good feedback on the incubating conditions by examining the newly hatched chick and how it hatched. A normal chick is one that hatched on time and without any help.

For this particular hatch, the operator (Patricia) noted that some chicks had a hard time pipping out of their shells and several of them were "dead in shell", that is, they were developed but dead. This tells us something was not perfect with the temperature, humidity, egg turning, or perhaps an infection. It could be a combination of many factors.

Monitoring the Incubating Eggs

There are two main ways to monitor the progress. Our preferred one is first by candling, and second by weight loss.

Candling is much more than just finding out if the egg is fertile. It helps monitor the actual growth of the embryo. Candling eggs is done by simply holding an egg up to a strong light in such a way that the rays shine through the shell to reveal the contents of the egg. In days of old, a candle was the light source, hence the term "candling eggs".

In our small operation, we use a modified night light with PVC L arm over it to direct the light to a right angle. The night light is connected to an extension cord. The disadvantage of this is that it gets hot so we don't use it continuously for very long. If you really need a strong light, a slide projector will light up an egg like a search light. We work quickly and are careful not to let the eggs get chilled by being out of the incubator too long, and we don't candle any one egg too long lest it get hot.

Throughout the incubation cycle we track how our eggs are doing. Often we have a shortage of incubator space and one way to make more room for more eggs is to candle and cull the infertile eggs. We first candle at 10 days, which is when the embryo is developed enough to see through the shell.

A fresh or infertile egg lights up like a golden globe. As the egg ages, an air sac forms because of evaporation. When handling an egg, we are careful not to rotate it too quickly, as this can rip the delicate membranes and cause damage to the embryo.

Home Made Candler

PVC elbow connector taped to night light with switch and an extension cord.

We look for vein development and air space size. When the embryo is approximately half way through the incubation period, there will be complete vein growth covering all but the air space in the egg.

Weight Loss Monitoring

Weight loss and the amount of air sac development is the best way to measure if the humidity levels are correct. We combine candling with the weight loss method of monitoring. Weight loss takes more time and attention, but provides a valuable insight into what is happening with the hatch that you cannot achieve through candling alone.

During incubation, eggs lose about 15 to 20% of their weight from the time of laying to pipping (pecking out of the egg). Ideally, this weight loss is consistent and easily plotted on a graph.

Weight loss is controlled by raising or lowering the humidity in your incubator. If eggs are losing too much weight, then you increase the humidity, which slows evaporation from the shells. Likewise, if losing too little weight, decrease the humidity, which will cause more evaporation from the eggs.

An Easy Cure for Straddle Leg

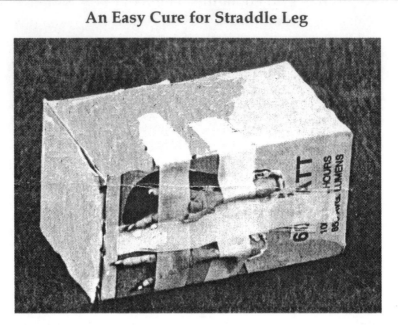

This might seem like chick torture, but by fixing the birds legs in this position for about 24 hours, you will enable it to walk normally. Also make a space in the top or end of the box for the chick to poke its head out to eat or drink. Any small box will do, we also use tea boxes.

The Hatch – Here They Come!

Three days before due date, we prepare our hatcher for the birthing event. We move eggs to the hatching trays and lay them on their sides. This allows the chick to exit the shell easier. Be careful not to crowd the eggs in the hatching trays so that the babies have room to push out of their shells. We candle the eggs as we move them from the incubator and discard the nonviable ones.

When hatching small birds such as quail we put covers on the trays to keep the babies from getting all over the incubator. We leave the hatching tray covers off for chicken and turkey hatches and have not had any problems.

Humidity is higher than for incubation by about 2 degrees, or 88 to 90 degrees Fahrenheit on the hygrometer. You can increase the humidity by adding another pan of water, decreasing the temperature by 1 degree, and/or adjusting the ventilation flow.

If eggs begin hatching a day or two early, then increase the temperature by about 1/2 degree and leave it higher for the next setting. If your eggs hatch a day or two late, then decrease the temperature by 1/2 degree for the next batch.

Causes of Low Hatch Rates

1. Infertile eggs.

2. Old eggs.

3. Parents stressed, unhealthy, nutritionally incomplete diet

4. Poor handling of eggs before setting for incubation (temperature, rough jostling, or unclean).

5. Shell contamination from dirty eggs, cracks, or thin shells.

6. Eggs not turned frequently enough.

7. Temperature too high or too low.

8. Too little or too much humidity in the incubator.

9. Improper ventilation or dusty incubator room.

10. Oxygen starvation – too much carbon dioxide built up in incubator.

Suggestions for Healthy Hatching

1. Set only fresh eggs less than a week old.

2. Follow your incubator instructions exactly.

3. Maintain the best sanitation for parents, nests, eggs, and incubator.

4. When storing eggs, maintain temperature at 55 degrees, (with a maximum range of 40 to 60 degrees) and wrap in plastic.

5. Maintain healthy breeding stock with correct hen-to-rooster ratios.

Problems with the Hatch

The four most common problems we had with hatching are listed below, followed by a simple table of common problems that give some reasons for specific problems at the hatch and during incubation.

1. Hard for chick to get out of shell; chick feels sticky. This usually indicates the hatch is not humid enough. The chick is having a hard time rotating inside the egg to pip around the blunt end and is drying out too quickly. The chick must be able to rotate freely inside the shell. This could also be an indication of humidity too low during part or all of the incubation period.

2. Chick seems too large for egg and appears bloated or swollen. This indicates the humidity might have been too high during incubation and the chick was unable to lose enough water. Even if an egg loses too much or too little water during incubation it will still need a higher humidity to hatch once it has internally begun pipping.

3. Early hatch. The temperature was probably too high. Check and recalibrate your thermometer, if necessary.

4. Late hatch. The incubation temperature was too low. If an egg gets temporarily chilled anytime during incubation this might cause it to hatch a day or two late. In these cases we might give a bit of help to get the chick out of the egg.

5. Straddle (splayed) leg & how to cure. We have had several cases of straddle leg and cured them all successfully. It is not genetic, but rather often caused by slippery surfaces or getting out of the egg. The leg ligaments get stretched such that the chick or poult is literally doing the splits with its legs wide apart. The bird cannot keep its legs underneath itself to walk. Straddle leg can happen to any species of bird.

Because of their long legs, turkeys and guineas are more vulnerable to straddle leg. This treatment is efficient, effective, easy, and costs practically nothing. We have used it with a 100% cure rate on turkeys, chickens, and guineas. Some of these were valuable heritage birds that we would have otherwise lost.

This is a treatment you must do as soon as possible, otherwise the leg ligaments are permanently stretched and the bird will never be able to walk correctly. The original idea is from Dennis Fett in his book *The Wacky World of Peafowl* – and it was pointed out to us in a workshop given by Herman Beck-Chenoweth. The idea is to put the bird in a natural, squatting, fixed-position for 24 to 48 hours. This gives the leg ligaments time to harden in the correct position which allows it to walk correctly.

Find a small box, like a tea or light bulb box. About 2/3 from the end of the box, cut 2 small holes in the bottom big enough to get the tiny legs and feet through. Position the legs so that the hock and feet are in a straight line and the two legs are parallel. This leaves the leg at the first joint (hock) bent causing the bird to squat (see photo on page 201).

Secure the legs in place with an easily removable tape. We use micro-pore paper tape that doesn't leave any sticky residue behind and comes off easily.

Make another opening at the end, or on top for the chick's head to come up, to be able to poke through to eat or drink. We put some cotton around the chick for additional body support and so the legs don't rub against the cardboard.

Now, instead of "chicken in a basket", you have "chicken in a box". Put the chick and box in the incubator or brooder and position it to be able to access water or food. Force feed it, if you think that is necessary. After about 24 hours, we let the chick try its legs. If it still has any sign of straddle leg we put it back for more physical therapy. After 48 hours, most babies are walking normally just like their hatch mates.

While you are taping the chick's legs to the box it might seem like you are torturing the little guy, but if you don't do this he will be condemned as a cripple.

If you have non-slippery surfaces that provide good footing, and you are still seeing straddle leg, and perhaps crooked beaks, then it might be due to malnourished parents. Make sure their diet has proper amounts of calcium (free choice oyster shell), phosphorous, and Vitamin D. Brewer's yeast and fishmeal are also good additions if you are having leg problems.

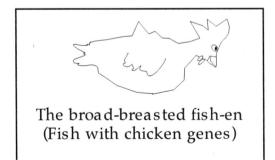

The broad-breasted fish-en
(Fish with chicken genes)

Genetic engineering gone astray

Problems when Incubating Eggs and Possible Solutions

Problem	Causes	Possible Solutions
Clear Eggs	Infertile eggs Chilled eggs	Get better parent stock or rooster/hen ratio. Feed parents better diet. Power outtage on incubator.
Dead in shell	Humidity and/or temp. too high or low. Temperature fluctuated. Incorrect turning, infectious disease	Better incubator monitoring and egg management.
Dead in shell with unabsorbed yolk	Too high humidity at hatch. Beak deformed prevents pipping.	Better parent stock. Heredity disease, parents malnourished or inbred.
Early hatch	Temperature too high.	Lower temperature. Calibrate therometer.
Late hatch	Temperature and/or humidity too low	Better incubator monitoring.
Sticky chicks	Too low humidity in incubator and/or hatcher. Too high temperature.	Better incubator monitoring.
Bloated, soft bodied chicks	Too high humidiy.	Lower humidity with better monitoring.
Weak, sickly chicks, low hatch rate	Too high temperature. Deficiency of amino acids. Weak or inbred parents.	Feed parent stock better, especially add soy beans, B-complex and trace minerals.
Early embryo death with blood ring	Chilled, eggs too old, too high or fluctuating temperatures.	Better incubator monitoring and egg management.
Embryo dying after 40% of incubation period	Infectious disease, incorrect turning, temperature too high or too low.	Better incubator monitoring and egg management.
Crooked toes or curly toe paralysis	Temperature too high or low. Inbreeding. Parents malnourished.	Adjust temperature, calibrate thermometer. Avoid inbreeding. Feed parents B2 and brewers yeast.
Crazy chick syndrome	Parents malnourished, especially lacking Vitamin E, B12, and trace minerals.	Balance diet, supplement with wheat germ.
Bent neck	Took too long to hatch.	Temperature and/or humidity too low in hatcher.
Straddle (Splayed) legs	Slippery surface in hatcher. Not enough calcium, phosphorous, or Vitamin D in parents diet.	Use box for leg adjustment (see text). Feed parents oyster shell, fish oils (meal), and brewer's yeast.

Chapter 10: Ordering & Brooding Chicks

This chapter explains how to determine the number of chicks or turkey poults you want to raise, how and when to place your order, and how to introduce your chicks into a brooder for their first few weeks of life. There is a comprehensive list of poultry suppliers in the reference section.

Broiler Chicks

Typical pasture broiler production yields a 4-pound dressed weight bird in 7 to 8 weeks. If you are ordering broilers for your own freezer, then you will probably want to order them all as one batch. If you are ordering broilers to sell to others, then you may want to split the order so that you will receive a manageable number of broiler chicks in each batch. Hatcheries set eggs based on the orders they have, and broiler eggs take 21 days to hatch, so you want to place your order at least a month before you want the chicks to arrive.

Upon arrival, the chicks will go into your brooder for 2 to 4 weeks, depending on how cold the time of year and how damp the weather. In the warm summer months, the typical brooder is set up to handle 100 chicks from one day to two weeks old. The chicks then go to the field shelter for an additional week of brooding before they are allowed to go out on grass. Generally, 1000 broilers require 1 acre of pasture.

Layer Chicks

Layers will start laying eggs in their 20th to 24th week after hatching. Each layer will generally lay 5 eggs per week for the first year, then after a two-month "molt" will lay 4 eggs per week, but generally larger eggs during the second year. Many poultry growers do not keep their hens for the second laying year, preferring instead to start a new batch from chicks each fall.

We time our layer chick purchases to 5 months before the farmers' markets open in our area, which is traditionally in April. This gives us a steady supply of eggs for our regular customers and the farmers' markets. The egg yield begins to decline with the short days of fall and winter, which coincides with the reduction in total sales once the farmers' markets close for the winter. Generally, 400 layers require 1 acre of pasture.

Turkey Poults

To keep turkeys warm during shipment, most suppliers require you to order a minimum of 20 poults. Turkeys are commercially available from about late March through July. To have turkeys ready for the Thanksgiving you would want to begin most of your poults in July. If your poults are born before July they are apt to be too large by Thanksgiving. We ask for poults to be hatched and delivered about the 3rd week of July.

How to Design and Build a Proper Floor Brooder

The long term health and welfare of your poultry flock depends greatly on how well they are kept during their first few days and weeks. If they are mismanaged during this critical early stage they will not perform well, and high mortality may result. As your operation grows, an electric brooder is well worth the investment. We found it was easier and cleaner to maintain, as well as being less stressful on the chicks.

Your chicks or poults need: fresh water, clean air and good ventilation, the right feed, plenty of warmth, and protection from drafts and predators.

Ideally a brooder is round, so chicks won't pile in a corner and smother. The sides should be high enough to keep out predators, and made of a material that will block drafts. We use 5-foot high wire fence panels covered with 6-mil polyethylene sheeting. The heat source can be a heat lamp with reflective shield,

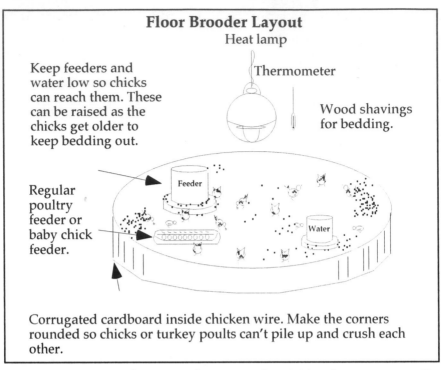

Floor Brooder Layout

Heat lamp

Keep feeders and water low so chicks can reach them. These can be raised as the chicks get older to keep bedding out.

Thermometer

Wood shavings for bedding.

Feeder

Regular poultry feeder or baby chick feeder.

Water

Corrugated cardboard inside chicken wire. Make the corners rounded so chicks or turkey poults can't pile up and crush each other.

or gas brooder with built in hover. It should be thermostatically controlled to maintain an even temperature. The 4-lamp electric brooder from Brower Equipment has a built in thermostat, and a supportive cage to hold the lamps the proper distance from the floor. (See resource guide).

Put 3 to 4 inches of bedding on the floor. A good litter is absorbent, light weight, of medium particle size, and provides good insulation. Some common bedding materials are wood shavings, ground corn cobs, and peat moss. The bedding insulates the floor for bird comfort and helps ease leg stress and disorders. Very coarse litter material can cause leg disorders. Fine litter, such as sawdust, can be too dusty.

Allow at least two to three weeks in the brooder for chicks, and eight weeks for turkey poults. Allow floor space of 1/2 square foot per chick and 1 square foot per turkey poult up to eight weeks. Start them in a smaller area and allow more room as needed. They grow fast and will quickly outgrow the initial brooder area.

How To Tell If There Is A Draft,Or If Chicks Are Scared

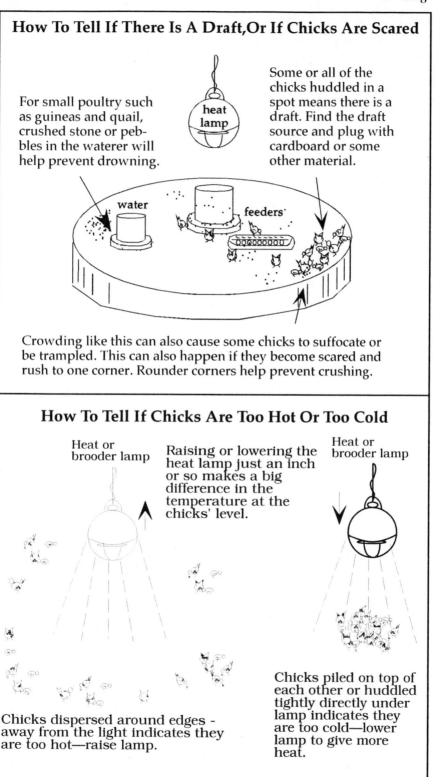

For small poultry such as guineas and quail, crushed stone or pebbles in the waterer will help prevent drowning.

heat lamp

Some or all of the chicks huddled in a spot means there is a draft. Find the draft source and plug with cardboard or some other material.

water   feeders

Crowding like this can also cause some chicks to suffocate or be trampled. This can also happen if they become scared and rush to one corner. Rounder corners help prevent crushing.

How To Tell If Chicks Are Too Hot Or Too Cold

Heat or brooder lamp

Raising or lowering the heat lamp just an inch or so makes a big difference in the temperature at the chicks' level.

Heat or brooder lamp

Chicks dispersed around edges - away from the light indicates they are too hot—raise lamp.

Chicks piled on top of each other or huddled tightly directly under lamp indicates they are too cold—lower lamp to give more heat.

The brooder temperature should be at least 90 degrees in the beginning and lowered by about 5 degrees each week. Don't put anything inside the brooder except poultry, water, feed, bedding, and a heat source.

Predators include bears, foxes, coyotes, possums, raccoons, weasels, mink, snakes, rats, dogs, cats, wolverines, owls, and hawks. Keep them out of your brooder by placing it in a protected area such as a barn or shed, and enclosing it tightly with good fencing.

Rats will be the hardest to keep out of your brooder and can do terrible damage while also eating your feed. Rats require vigilance, early detection, and a good trapping plan. A friend recently lost 36 broiler chicks in one night to rats. The rats tunneled into his brooder under the floor. A rat can eat almost as much grain as a hen. So even for economical reasons, keep the rats out if possible. Whenever we see a rat we immediately set traps and are very aggressive about clearing all of them out. Even in the best of systems, rats happen.

Clean the brooder thoroughly between batches, and always put clean bedding. Buy planer shavings for bedding. Enough planer shavings for 100 chicks costs less than $4. Sawdust is too fine, and the birds will try to eat it, which leads to compacted digestive tract. Don't use hay or straw, they mat quickly and will not absorb wet manure.

Battery Brooders

We bought a 5 level battery brooder and it made a difference in our operation. We could easily care for up to 500 chicks in a room approximately 12'x12'. Battery brooders are easier to clean than floor brooders. Manure drops through a screen to a bottom tray that has an absorbent liner. This kept the chicks cleaner and dryer. Every day or so we would remove the bottom trays and put the liner and manure into the compost pile.

Five Level Battery Brooder

With this battery brooder we can keep up to 500 chicks at a time. Note the tracking cards clipped with each batch. Feed trays are along the sides and water troughs are on the ends. Each level has a heat source. This unit is on wheels which helped manuver the unit when servicing the chicks and poults.

We had our brooder in an insulated room that kept drafts to a minimum and let us regulate the temperature very precisely. Sheds and garages usually don't make very good battery brooder rooms because of the lack of environmental control.

The chicks give off a lot of humidity and carbon dioxide so good ventilation is essential.

Our brooder room has a concrete floor so we can sweep up any feed that has dropped, and spill water without causing a mess. Because this method doesn't require bedding, such as wood chips, there is much less dust generated with battery brooders than floor brooders. With screens on the door and windows we can also keep flies somewhat under control.

We keep chicks in the brooder for up to two weeks. If it is cold or rainy, we would transfer them to a temporary floor brooder until the weather improves, and we might keep them in the brooder another week to save them from being stressed.

Each level of the brooder has its own heat control so we can decrease the heat as the chicks grew.

One problem we had was with the chicks' and poults' tiny little legs getting caught in the wire mesh screen. This was especially true with some of the baby turkeys. They would get their long legs caught below the hock in the mesh screen and could not get back up. The legs would get swollen. In these cases, we had a smaller floor brooder that we would put them into to recover for a few days and they were usually fine.

Receiving and Caring for Chicks and Poults

These little babies are living beings and it is your responsibility to give them what they need to survive and thrive. During the first few weeks, attention and proper care are especially critical. Always have your brooder ready at least one day before your chicks are scheduled to arrive.

We ask our local post office to call us when chicks arrive so we can pick them up sooner than the normal delivery. This helps decrease the stress on the chicks. The post office would usually call about 5:30 am, which only partly explains Andy's "deer-in-the-headlights" look.

When you order your chicks or poults, the hatchery will give you a shipping date. They will ship your order via Priority Mail in the US Mail. Neither United Parcel Service or Federal Express will ship live poultry. Some things can go wrong in shipment. The shipment could get lost and wind up sitting for hours or even days on a loading dock. The airline may refuse to carry live birds if the temperature drops below 45°F or goes above 85°F. Your local post office may fail to notify you when the birds arrive.

Hopefully, you will find a reliable hatchery close enough that birds are delivered by surface mail and don't have to be subjected to airline transportation at all.

Chicks arrive in a cardboard box with ventilation holes. A typical box has 4 compartments. Each compartment can hold 25 chicks or 20 poults. The body heat from this many birds will keep the temperature up so they don't become chilled and die. In the warmer summer months, the hatchery may choose to place fewer birds per compartment, thus regulating the interior temperature of the box so the birds don't overheat.

Chicks and poults can survive for up to 3 days after hatching without food or water, but the sooner you can get them unpacked and place water and food in front of them the better. Hours can make a difference in the stress of these baby birds.

As you count them out of the box and into the brooder, dip their beaks in water and then in the feed so they will learn quickly. To help introduce them even better, put a newspaper down with some grit and feed on top. They will immediately begin pecking the food and they need the grit for digestion. Within hours they will all know where the food and water is.

Turkey poults are more fragile than chicks and are almost blind as babies. They are attracted to bright things so one of the tricks we use to help them find their food is to put marbles in the feed. They will be attracted to the marbles and the feed at the same time.

Observe the flock to make sure the brooder heat source is set properly. If all the birds are piled directly under the heat source, then it is too cold. If they all scatter away from the heat source, then it is too hot. Ideally when they all bed down for the evening they will be ringed around the perimeter of the heat source area.

If they are all assembled at one side of the brooder, it may indicate a draft. It can also mean they are being preyed on by a rat and are assembled away from the area where the rat has burrowed in.

Chick Feed & Grit

We use a commercial chick starter for at least the first two weeks. If the chicks are having trouble recognizing the starter mash, then just cover the bedding near the feeders with a layer of newspaper and spread some feed and grit on it. This helps them find the food. The newspaper will be needed only for a day or two.

Grit is available from most feed stores. Think of grit as the bird's teeth – it helps the bird grind up its food. You can buy grit as limestone, oyster shell, or granite. We use granite for our chicks. Sprinkle the grit around like you would salt. Don't put down too much or they will fill up on grit instead of food.

Have at least two feet of feeder for every 25 chicks. Otherwise some of the more timid chicks won't get enough to eat. Check them at least twice a day and keep the waters and feeders clean and full.

Never feed layer ration to baby chicks or poults. The calcium is too high and can harm their kidneys.

Chicken Whispering

We listen to our chicks and they tell us a lot. When you approach the brooder, the chicks should be making peaceful sounds and are not alarmed by your presence.

If they are cheeping shrilly and scurry away from you it indicates they are being preyed upon, probably by a rat. They will also cheep shrilly when they are making their needs known, such as when it's too hot or too cold, or when the lights go out unexpectedly, or when they are hungry or thirsty and the feeder or water fountain are empty. Occasionally a chick or poult may escape from the brooder and will cheep shrilly and try to get back in with its flock.

Always listen for cheeping that is out of the ordinary. They will tell you what they need if you only pause, be aware, and listen.

First sketch of the Day Range Cover with Atilla the Hen as our model. Atilla thinks the rhinestone glasses make her look smarter, and the bonnet puts her on the cutting edge of farm fashion.

Chapter 11: Feeds, Feeding, and Water

> No feed ration is set in stone and advances in nutrition are made almost every day. We keep changing our rations given new information and ingredient availability.

When calculating your feed budget, here are some handy rule-of-thumb guidelines to use:

- A broiler eats about 15-16 pounds of feed from day 1 to day 56.

- A layer eats about 100 pounds of feed from birth through the end of her first laying cycle.

- Broad Breasted White and Bronze Turkeys eat about 60 pounds of feed from birth to harvest weight at 15 or 16 weeks old.

We have suggested several formulas for rations in this chapter, but advances are always being made in nutrition. We changed our formulas very often depending upon what ingredients were available and the latest supplement that we read a favorable report on. We keep learning how to do better and better.

One resource we highly recommend you contact is the Fertrell company in Pennsylvania – talk with Jeff Mattocks. Jeff, or someone in his office, can give you the most recent Fertrell ration recommendations. Fertrell (800-347-1566). This can save you a lot of guessing and money and can enhance the health of your birds.

If you grow several hundred birds, it will be worth your while to find a local feed mill to grind and mix a custom feed ration for you. They can bag it, or their feed delivery truck can put the feed directly into your feed bin. For us, it is much easier to get bulk deliveries to our grain bin and then to use 5-gallon buckets to divvy the feed out. This also has the advantage of not creating feed bag waste.

We calculated that the savings in getting bulk versus bagged feed paid for our 4.4 ton feed bin in the first year. A new feed bin costs about $1,200 installed on your concrete pad.

If you are only growing poultry for yourself and a few to sell, you may not need enough feed to warrant having a ton or more ground and mixed. In that case, just simply buy feed from your local feed store.

When buying feed, ask for 20% protein for your broiler starter, and 28% protein for turkey starter. Turkeys need 28% starter for 4 weeks, then 22% grower for 4 weeks, and then drop to 18% grower/finisher (use broiler ration) for the remainder. Broilers need 18% grower/finisher, and layers need 16% protein feed. In areas where turkey starter is not available, you can make do with "game bird starter".

We have our feed mixed at the local mill, and we provide the supplements that the mill doesn't stock. The supplements we use are Poultry Nutri-balancer and Aragonite, both available from Fertrell (See Resource Guide).

Aragonite is a more natural replacement for lime, and Nutri-balancer contains the vitamins and enzymes so important for peak health and performance. Here are typical Fertrell recipes for one ton of feed:

Pasture Layers 16% protein:

Cracked corn	1,000 pounds
Ground roasted soybeans	600
Whole oats/spelts	200
Poultry Nutribalancer	60 (available from Fertrell)
Kelp meal	16
Salt	5
Aragonite (Fertrell)	140
(or use 1/2 Lime and 1/2 oyster shell)	

====

about 1 ton or 2,021 pounds

Pasture Broilers 18% protein:

Cracked corn	700 pounds
Ground corn	700 pounds
Soy bean meal	450
Sealac, fish meal	50
Aragonite/oyster shell	25
Poultry Nutribalancer	60
Soybean oil	25
	===

about 1 ton or 2,010 pounds

Pasture Broilers Starter 20% protein:

Cracked corn	1,000
Roasted soybeans	622
Oats/spelts	200
SeaLac fishmeal	75
Poultry Nutribalancer	60
Aragonite	25
Natural mineral salt	5
Kelp meal	15
Fertrell Probiotic	2.5
	===

about 1 ton or 2005 pounds

Turkey Starter 28% protein:

Cracked corn	300
Spelt or oats	500
Ground roasted soybeans	1,000
Sealac fish meal	100
Aragonite	15
Poultry Nutribalancer	80
Natural mineral salt	5
Dicalcium phosphate	8
	===

about 1 ton or 2,008 pounds

Turkey Grower 22% protein:

Shelled corn	1,152
Soybean meal	600
Sealac fish meal	100
Aragonite/oyster meal	25
Poultry Nutribalancer	60
Soybean oil	50
Kelp meal	10
Probiotic	2.5
	==

about 1 ton or 2,000 pounds

Remember, these are suggested recipes for ingredients that are usually available in our area. You can vary the amounts and feedstuffs depending on availability and price.

Your feed mill will work with you to calculate and adjust the amount of protein required for each ration. There are also different nutritional supplements on the market, or that occur naturally that you might experiment with.

Nutritional Supplements

We try to get the best nutrition for ourselves and our animals, but not necessarily the cheapest. We are especially concerned about the availability of trace minerals, the quality of the fine oils, and proper amounts of vitamins for optimal health.

We feel the soils on which most of the grains are grown are nutritionally and biologically dead. Harsh fertilizers and tilling methods have killed most of the flora and fauna (including earth worms) in the soil. Additionally, since NPK (Nitrogen Phosphorus Potassium) and ammonia seem to be the primary fertilizers, soils, with all the erosion, have lost many of the trace minerals and amino acids that are essential for good health. This lack of nutrients follows right up the food chain. An enduring classic on how what we eat shapes us, for better or worse, is *Nutrition and Physical Degeneration* by Dr. Weston Price.

We believe that a superior ration gives us a profitable payback in healthier animals with better immune systems. In the long run, we spend more on feed, but our vet bills are lower, our animals have a higher resistance to diseases, fewer parasites, better fertility, and more efficient feed conversion. The bottom line is we have very low losses due to morbidity and mortality.

Of course we are always on the look out for when more supplements might be needed. For example, if we begin to see leg problems, then we add additional brewer's yeast or distiller's grains to the ration. These are high in the B vitamin complex and help prevent leg disorders. We also use homeopathy to treat certain disorders and have found it to be highly effective. Below are a few of the supplements we routinely use.

Comfrey. A feed supplement you can grow yourself. Comfrey is rich in vitamins and minerals and has many other beneficial effects as a forage. It also contains allantoin which has the ability to aid in cell regeneration, wound healing and cell proliferation. Doing this helps the immune system protect the animal against infective diseases. Allantoin is the primary ingredient in products such as canker medicines. We talk about comfrey in detail in our book *Chicken Tractor.*

Duckweed. Another nutritious supplement that you can grow yourself. It is considered a problem weed in some ponds. However, duckweed can produce up to 20 tons of mass per acre per year. It has a protein content of about 18%. Compare that with alfalfa which yields about five tons per acre.

Duckweed will grow on any pond surface and won't interfere with your irrigation or livestock watering. Just skim off enough duckweed each day to supplement your chicken feed. The duckweed prolifically reproduces giving you a steady supply all summer until it goes dormant in the winter.

It comes back to life in the spring and keeps on producing. Our hope is to eventually replace as much as 20% of our feed bill with home grown duckweed. We feed it to all the livestock, including the pigs, goats, and poultry. Duckweed can also be dried for future use.

ProBiotics. These directly fed microbes help establish the beneficial bacteria in the gut which aids absorption of nutrients, increases feed conversion and enhances the immune system. We add one three pound bag (cost about $5) to each ton of ration. The good bacteria include Lactobacillus Acidophilus, Bacillus Licheniformis, Bacillus Subtilis, and Bifidia. These are similar bacteria to what is in organic live culture yogurts, which we enjoy eating.

Salt. Often we add salt to our feed mixture, especially when it's hot. We prefer the natural formation unprocessed salt that contains many valuable trace minerals and no fillers or binders. For the hoofed livestock, we get the salt in chunks and make it available free choice just the same as the commercial salt blocks. The natural chunk salt is more weather resistant than the pressed blocks and there is less waste.

For poultry, we use finely crushed salt and have the feed mill add it to the ration. One source of natural rock salt is Redmond Minerals, in Utah. See the nutritional supplements resource section on how to find where their products are sold.

Sea kelp. One of our favorite supplements for providing over 60 trace minerals and vitamins. We add kelp at the rate of about 1 to 2 percent of the feed. We also mix kelp with salt and or diatomaceous earth and let the animals free choice. Unfortunately, you probably can't grow your own kelp but you can buy it from your feed supply or by mail. We get ours in 50 pound bags and feed it to all our livestock, including our dogs and cats. It costs about $30 for a fifty pound bag.

These two pages are part of a tri-fold promotional flyer we use to educate our customers on nutrition. Feel free to use any of the copy for your family farm advertising to help promote your products.

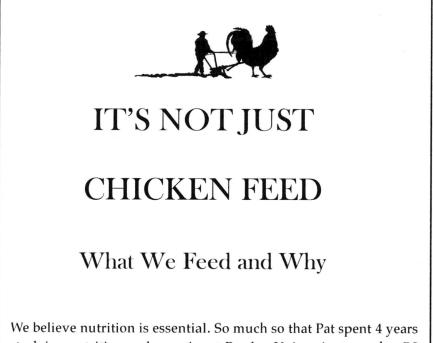

IT'S NOT JUST

CHICKEN FEED

What We Feed and Why

We believe nutrition is essential. So much so that Pat spent 4 years studying nutrition and genetics at Purdue University to get her BS in agriculture along with her degree in Pharmacy. We know well the role nutrition plays in an animal's health, and that nutrition (or lack of) flows directly up the food chain.

We give our animals the best nutrition we can provide - not the cheapest. We feed our birds a ration that is as wholesome as we know how to make - we often even see our dogs and cats munching on it as we are feeding the poultry. We have our ration custom mixed at a local grainery and have it delivered in 3 ton batches - enough for about 3 weeks. That way our birds always have a fresh ration.

Of course our livestock are also on organic pasture that is free of pesticides, herbicides, and harsh chemical fertilizers. This provides the essential sunshine, fresh air, and nature's finest bugs, weeds, and seeds to supplement their diet.

Here is a summary of what we feed our birds:

Local Grains – as much as possible we use grains that are produced withing the Shenandoah Valley. This helps our local farmers and economy to keep money within the community.

Corn – course ground and fine cracked. This gives the birds primarily energy. The texture of the corn helps provide roughage and variety to the ration.

Hi Protein Soybean Meal – primarily for protein and high quality essential oils.

Roasted Soybeans – roasting the beans helps preserve the oils which are extremely high in vitamins and amino acids.

Oats - for carbohydrates and vitamins.

SeaLac Fish Meal - this is about 60% protein and adds a high concentration of vitamins and trace minerals to the ration.

Brewer's Yeast – loaded with the B complex vitamins that help keep leg problems to a minimum and boosts their immune system.

ProBiotics – direct-fed microbials from Fertrell. These are the good bacteria such as are found in organic yogurt. The probiotics stabilize the entire digestive tract of animals and help with the fermentation that is necessary for complete digestion and optimal health. These live cultures increase the absorption of proteins, starches and fats. We feed this to all our livestock and even use it ourselves.

Sea Kelp – from Iceland that is loaded with trace minerals and vitamins so numerous it could only have come from the sea.

Aragonite – high quality calcium suppliment that also provides iron, and trace minerals.

WHAT IS NOT IN OUR FEED

- No antibiotics
- No hormones
- No artificial food additives or preservatives
- No animal by-products, toxins or fillers

We appreciate your business!

Water and Dealing with Coccidiosis

By Tim Shell

There are many growers who have not understood the necessity of good water, and the potential ravages of coccidia in pastured poultry. They can cost the pastured poultry producer much in the course of a year. The degree of loss is proportionate to the degree of uncontrolled infection. Minor infections can cause poor growth and lower dress-out weights. Major infections can cause significant mortality. In discussing pastured poultry with many folks around the country it has become clear that many pastured poultry farmers are suffering major and minor losses in flocks from coccidia. Growers profits are at risk if they are unaware of the symptoms and the management style needed to avoid losses due to coccidia.

Coccidia are everywhere. They are carried in the feces of almost all wild birds. They are on your farm. Coccidia is a protozoa and has many strains. The coccidia that infect chickens does not affect other livestock. Always assume your poultry is exposed to coccidia. If well cared for, most birds develop an immunity to the parasite early in life. Exposure is not the problem. Exposure to pathogens is critical to the vitality of an immune system.

Overexposure and extended duration of high stress events are what create the conditions that allow the coccidia to overwhelm the bird's system and cause losses. Losses from coccidia are generally the result of a breakdown in the poultry person's management which results in overexposure of the bird to the organism. If field conditions are highly stressful for extended periods of time the birds immune system may be weakened to the point where it cannot control the organism and setting the stage for the development of coccidiosis.

There will always be a percentage of our livestock that never cease to amaze us with the extremes they can be pushed by our bad management and still survive (shame on us). On the other hand there will always be the few that have a predisposition to get sick in any group of livestock even when the best of management has occurred and optimum conditions are provided. Nature normally dispatches these creatures to prevent passing on the tendency to future generations. It is generally unprofitable to keep what nature has discarded. Our job is to provide optimum conditions in a model that is easy for us to manage. The easier a job is to do the more likely it is that it will get done and be well done. Lets review some of the basics of raising poultry.

1. Clean Water

Would you be willing to drink the water you are forcing your birds to drink? If the answer is no then their water is not clean enough. Go out to your poultry and look at their water and suggest to yourself that you take a drink. The degree of aversion you have to taking a sip of the birds water is proportionate to the degree of risk of excessive infection the birds are faced with. The greater the aversion, the greater the risk.

Don't drink it, just use your reaction to judge the state of the water. Most people with coccidia problems would get over them right away by following this clean water rule. You tend to loose contact with exactly how dirty the water is until you shock yourself into reality by suggesting you take a drink. "Yuk! It's dirty!" The birds opinion about it is mutual. Put out clean water and see which one they choose. Especially for young stock, you must not allow for dirty water.

The big offenders here are brooder bedding (which contains feces) being scratched into the waterer or fecal/bedding dust collecting in/on the waterer or ground puddles in rainy weather. I would attribute the source of most losses related to coccidia to the last week in the brooder with dirty waterers — I'm-going-to-put-the-birds-out-on-pasture-tomorrow syndrome.

The best poultry growers have the cleanest waterers. Clean your waterers regularly. Be sure to eliminate any wet or dirty places/puddles that the birds might have access to. They do not understand not to poop in (coccidia out of the bird, into the puddle) and drink out (coccidia out of the puddle, into the bird) of the same puddle. Mud is dangerous if they have access to the outdoors.

2. Clean Bedding

Be sure the birds have bedding to lounge in clean enough that you would be willing to kneel in it wearing nice work clothes and show the peeps to your best friend's children.

Wet, dirty bedding causes a hygiene overload for birds which can overtake their immune system. All livestock have specific tolerance thresholds for specific pathogens. Above that level they get sick. Below that level they do not get sick. Exposure to a disease causing organism below a certain level of colony forming units does not cause disease even though the pathogen is in the birds system and you never know it because they are dealing with it as planned.

The fact that many of us have gotten away with allowing some horribly dirty conditions for our poultry is not a tribute to our skills as poultry persons but to the wonder of the birds incredible immune system. It does not mean the birds were never exposed to germs but that their immune system was successful in overcoming the invasion.

Poultry exhibit hygiene behavior that includes sliding their beak over their feathers to remove dirt. Their system is designed to handle ingesting a certain amount of dirt each day. If you give them more than they can handle they may get sick. If the dirt is from overly soiled bedding they will be ingesting their feces via preening. You will notice that the coccidia affected birds will look dirty.

They "know" somehow to stop preening until their body can handle the crud. Your other birds would look that dirty too, they just clean up every day. The amount of dirt on the unpreened ones lets you know how much the others are cleaning off themselves in a short time. You must add bedding in whatever amount needed to deal with problem spots in the brooder like around the waterers and in the nightly sleeping spot. Also if some of the chicks are at a size disadvantage they may get walked on if they pile up to sleep on cold nights and have more than their fair share of dirt to clean off themselves the next day.

3. The Double Whammy — Out of Food and Water

The busy person easily falls prey sooner or later to the fault of letting the feed and/or water run out for their poultry. The combined effect of hunger and/or thirst on the birds is to encourage them to sort through the ground/bedding searching for particles of food or moisture.

In a dirty environment, hungry and/or thirsty birds will almost certainly exceed the safe threshold of tolerance for pathogens as they ingest soiled material in search of food and water. This exposure level, coupled with the stress caused on their system by the hunger and thirst creates a situation ripe for disease to set in. If you are very busy it is easy to miscalculate the time your stock have been without food and or water and assume they are not too badly affected by it. But the health of the bird is the sum of the care for each of its needs that it has been given in its life.

4. Sunlight and Coccidia

Poultry love sunlight. They love to sunbathe. This is a great benefit to them. If the weather turns cloudy and damp or the birds have no access to direct sunlight in the brooder at early stages of life they may be at a disadvantage for proper hygiene. The sunlight is a disinfectant and a therapeutic tonic in the bird's world. Doing without can contribute to outbreaks of coccidia as well as other diseases. You should have abundant natural daylight in your brooder. The combined affect of violating all of the above principles can be disastrous.

Symptoms of Coccidia

Birds faced with an overwhelming infection of coccidia will look dirty and unkempt. They will be weak and listless, hunkered down in a corner and not moving much. They do not look healthy one day and just drop dead the next. You can tell several days ahead which ones are on the way out. They can have bloody manure from the bleeding of the intestine caused by the irritation of the coccidia on the papillae. Severe infections will have foamy, yellow, mustard like manure. If you have birds in this condition you have already experienced significant losses in the productivity of the rest of the flock.

Left untreated coccidia can lead to necrotic enteritis (followed by death) which is the sloughing off of the inner lining of the intestine which is where the coccidia take up residence and cause intestinal bleeding. Birds suffering to this degree should be put down as recovery is not likely.

Treatment of Coccidia

1. Create an environment with abundant natural sunlight.

2. Deal with dirty bedding and water. Use a plywood circle under the waterer large enough to keep the birds from scratching feces into it. Elevate the platform 3 to 4 inches above the bedding. Use drink cups or nipple waterers to provide sanitary water.

3. Eliminate the wet pack areas around the waterer by removal and rebedding. This wet, soiled area is highly conducive to the exponential proliferation of anaerobic pathogens such as salmonella and E. coli. If you have had a severe infection in

the brooder clean it out and disinfect and re-bed with clean product.

4. Supplement with water based probiotics in the waterer. Available from the Fertrell Co. I had a severe infection in my flock last year due to management failures. Jeff Mattocks of Fertrell recommended fresh raw cow or goat milk to me as a supplement in or on the feed or fed free choice as a successful remedy for coccidiosis.

The milk is mucus forming and coats the intestinal track. It also has beneficial bacteria and enzymes in the raw form. I used it on the feed about 2.5 Gal milk/5 Gal Bucket (25 to 30 lb) of feed, well mixed. Putting it in the feed makes sure the birds all get a dose. Usually the birds turn around in 48 hours. An old poultryman told me they used to use milk products to treat coccidia before the medications came out.

The routine use of coccidiastats in commercial poultry feed is indicative of a model that forces the birds to live in the presence of their own feces to an abnormally high degree. There are models, thankfully, that have solved the problem by reducing the exposure to fecal material to a level that does not cross the birds' threshold of tolerance for coccidia. Most of us have opted for those better models. Lets make sure they are operated correctly.

Chapter 12: Pasture & Forage Management

"If vain our toil,
we ought to lame the culture,
not the soil."
Pope — Essay on Man

The pasture component is not well understood by today's farmers because it really isn't clear how much food value the poultry can extract from the forage. We do know that the broad breasted white and broad breasted bronze are just as aggressive at foraging as any of the minor breeds of turkeys we have tried. This is contrary to some reports that the large birds will fall apart on pasture but, in our several years experience this is simply not true.

Likewise, we see the Cornish Cross broilers as good foragers, maybe not as active as the old line dual purpose breeds, but certainly foraging enough that we can be satisfied. However, whether or not they can convert that forage to food value is not clear. Older reports that they can get up to 30% of their food from pasture are simply inaccurate.

Our farm in Virginia's Shenandoah Valley is 41 acres of rolling hills that are not well suited to grain or plow agriculture. The land is ideally suited to graze. None of the land has been fertilized or limed for at least two decades, and as a result the topsoil is pretty well devoid of nutrients.

To fertilize and establish good pasture, the USDA Cooperative Extension recommendations using herbicides to kill the existing growth, plow down the existing turf, disc harrow, fertilize and lime, and reseed. This costs about $300 per acre or more. To do our 30 acres of pasture and hay fields would cost us about $9,000. It's better for us to invest that $9,000 in an income earning farm enterprise such as turkeys.

We fence in an area, turn some turkeys on it, give them plenty of feed and water, and let them graze off whatever they can find in the worn-out pasture. If we leave the turkeys in one place long enough they will actually kill the existing growth. When we move them off, the dormant clover seeds will germinate and the resulting pasture will be rich in diversity and well fertilized. Rainfall on the manure forms carbonic acid which helps break down the natural limestone in the area and that increases the pH of the soil. We think limestone is so cheap, however, that it makes more sense for us to go ahead and apply it. We would apply at the rate of about 2 tons per acre, for a cost of $36 per acre.

At any rate, what we can do is earn an income from the birds while they are improving our pasture sward. If we have 100 turkeys per acre and make $20 net income per turkey, then we earn $2,000 per acre instead of having to spend $300 per acre as the USDA recommends.

Eventually we would like to get to the point on our farm where we can grow all the grain that we need to feed our flocks and then get the fertility to grow that grain from our flocks. It is sort of a perpetual motion machine, if you will. How would that look?

Let's assume 100 turkeys per acre will need good graze and about 60 pounds of feed each to make a good market size bird in a reasonable period of time. Our harvesting target is a 12 to 20 pound bird, since some customers want smaller and some want larger. We are using our own poults so that the ones born  early in the spring will be quite large by Thanksgiving, and the ones born later in the year will weigh a good deal less. Thus we need 6,000 pounds of grain to feed that flock. And, they can put down enough fertility to make a good grain crop possible.

According to USDA research in the 50's—back when almost all turkeys were grown on free range—each pound of feed on pasture will return 1.1 to 1.5 pounds of manure, of which 75% is water. Turkeys are not especially good at extracting available nutrients from their feed. The resulting manure is quite high in nitrogen (N), phosphorus (P), and potassium (K) as well as many trace elements, especially since we feed our flocks a wonderfully rich diet containing kelp, diatomaceous earth, aragonite, and Nutri-balancer from Fertrell. We also use locally grown wheat, oats, roasted soybeans, and corn.

The turkeys will weigh about 2 to 4 pounds when we put them on pasture at 4 to 8 weeks of age. One hundred turkeys per acre can put down about 60 pounds of manure each, or 3 tons per acre. Nutrient variability is extreme, but we figure the manure has a 10-10-10 value of NPK fertilized. Therefore we're putting down easily 240 pounds of nitrogen per acre on a dry weight basis.

This would be way too high for the following crop except that the manure is placed on the sward over an 8 to 12 week period, thus giving the soil microbes a chance to assimilate the manure slowly as it is applied. As is the case with many organic fertilizers, only about 50% of the available nutrients are released per year. Only if you leave the turkeys in one place too long will the manure overload become extreme. This occurs mostly in their roosting area since they will deposit a third of their daily manure on that small area.

To repeat, that's why it's important to use the fence, feeder, fountain, and shelter as the brakes, steering wheel, accelerator, and engine of the feeding system. The perimeter fence keeps the birds in and the predators out. We move the shelter weekly, and the feeders and fountains daily, so the turkeys are always traveling, grazing, and resting in a different part of their paddock.

What we found as we continued along the journey of discovery in the turkey world, was that many of the techniques we were using to raise good turkeys can be used for other livestock as well. We also use portable fencing, feeders, and fountains, for example, to guide the livestock where you want it to graze. Sometimes the feeders can be along the fence line so we can reach the tractor bucket across and deliver the feed.

We've learned to look at the electric fence as both the brake and steering wheel. The feeders, fountain, and shelter are another part of the steering wheel that helps guide the turkeys onto the areas we want them to graze, or manure, or both. Whatever our objectives are we can accomplish them simply by moving one or all of the furnishings that make up their pen.

A secondary benefit, of course, is that you can have a healthier, better-looking turkey to sell to your customers. Because of the variety in their diet, and the exercise they are getting on range you will have a much tastier bird that will justify the higher prices you are charging for your product.

Also, by rotating the birds around the paddock you can lay down an even layer of manure that is rich in nutrients from undigested feed and from the forage they are grazing. Poultry manure is rich in polysaccharides that help to glue soil particles together. This alleviates some erosion and makes soil more capillary.

The year following the turkey rotation you will have a richly fertilized sward. If you manage the sward carefully to encourage legumes you can have a forage high enough in protein to pasture lactating dairy cows profitably.

In our rotation we are planning on having the turkeys graze the same paddocks for two years straight. At the end of the second year we move the turkeys off and plow down the graze and accumulated manure and mulch it for the winter.

Then the following spring we can till the soil and plant market gardens or berry patches or whatever we want to do that can take advantage of the richness of the soil fertility supplied by the turkeys.

At our farm, we don't have a high-value livestock enterprise, such as dairy cows, to follow the turkeys, so we rely on the market garden or berry patches to extract the value of the stored nutrients. An added benefit of moving the turkeys off that particular site for the following two years is to rest the soil from turkey pressure and break up any long term parasite or disease problems that may be happening. The same can be done with grain crops for turkey feed.

One disadvantage of growing grains for turkey feed is the storage requirement from one year to the next. Grains harvested in the fall one year will have to be stored until fall of the next year with some resulting loss in nutritional quality and value.

In this scenario the only products the farmer will have to sell from these plots are hay, straw, broilers, and turkeys. It might also be worthwhile to look at including a ruminant such as dairy cows or feeder steers in the rotation to give the farmer additional products to sell.

For this to work it requires fenced paddocks, corn picker, grain combine, soybean roaster, and feed bins to hold the grain until the following year. Also we'll need a mixer grinder. Then we have to purchase our supplements such as sea-lac, nutri-balancer, aragonite, kelp, probiotics and diatomaceous earth. Plus, a baler to handle the straw and to handle the hay making. Also, we will need to have beef steers at the right stage to graze the grass/legume plot. We could also do this with dairy cattle at 2 cows per acre for example. It makes most sense, however, to pasture broilers on the new ground simply because it will have less impact on newly planted ground and the first lush can be taken off as hay.

No-Till Market Gardening

It is possible to have a no-till market garden system using turkeys. First, use a spading machine to work the soil into a useful seedbed. Then plant a crop and let it grow out to maturity, harvest it, then turn the turkeys in on the land. Let them hog it down to bare dirt in a few days, then move them to another location. At the garden site put down soil amendments such as lime and micro-nutrients that your soil test indicates are needed. Then mulch the whole site and let it sit through the winter.

In the spring, lay out beds with irrigation tapes and then use transplants to replant the garden. Just rake back mulch at each transplant site and place the seedling. As weeds poke out, just cover them with more mulch. At the end of the garden season, put the turkeys back in to hog it off, and add more mulch for the following winter.

In an ideal system, the poultry would be rotated with the pasture and grain crops in such a way as to make use of their fertility. It seems that poultry can lay down enough manure to fertilize a following grain crop. So each farm would then size its poultry flocks to meet the needs of fertilizing the grain crops and the grain crops would provide the feed ration required by next years poultry.

There are some difficult logistical concerns, however, including how much grain storage is available, and whether the farm can produce the right quantities of feed grains.

Poultry diet is made up primarily of corn, soybeans and perhaps wheat or oats or some other small grain. The soybeans have to be roasted before grinding. The grain needs to be held through the winter, so it must remain whole. Once grain is ground the nutrients leach away or spoil quickly.

In the last chapter we discussed the dietary needs of the poultry. Now let's consider just how much grain is required. If we are to raise 1,000 turkeys for example, we will need about ten acres to pasture them on. We will also need about 60,000 pounds of feed, or roughly 30,000 pounds of corn, 18,000 pounds of soybeans and 6,000 pounds of small grains. We can expect corn yields of 6,000 pounds per acre, so we'll need 5 acres of corn. We can expect soybean yields of

2,000 pounds per acre, so we need 9 acres of soybeans. We can expect 2,000 pounds per acre of small grains, so we will need 3 acres for small grains. Total land required is:

Pasture   10 acres
Corn   5 acres
Soybeans   9 acres
Small grain   3 acres
For a total of 27 acres to raise 1,000 turkeys.

What happens instead is that millions of acres of grain are planted in the midwest and the feed is then exported to other areas of the country where the poultry are grown.

The hilly land on our farm is totally unsuited to grain culture, so we can only rely on pasture culture. We must then search for grain growers in our area or region from whom we can buy the grains, then have our local mill grind them into feed, or figure out some way to do the grinding ourselves.

Harvesting and Storing Water

Permanent agriculture is about care of the earth and care of ourselves. All things that enter our farm are energy, some good and some not so good. We can learn to design to optimize the good and minimize the not so good. Permaculture and the 2nd law of thermodynamics both say that energy goes from a useful state to a useless state. Our job is to keep energy in the useful state for as long as possible.

Permaculture is similar to value added. By keeping energy on the site as long as possible we will ensure a positive return on our costs and labor. One way this can happen is with water. How does water enter our farm?

1. Run-off from adjoining farms. Use swales, spreader dams, check and vegetative buffers dams to slow water down so it can seep into the soil. Follow contour or off-contour to direct water to ponds.

2. Springs. Source of potable water for humans and livestock, cool water for trout and bass farming, feed ponds for recreation, aquaculture, water gardens, and market crops such as watercress as well as wildlife habitat.

3. Rainfall. Water harvesting, storage, and surface erosion control, refills ponds and cisterns and recharges the water table.

4. Streams and rivers. You can use ram and sling pumps to move water to storage places.

5. Ditch water in some parts of the country.

6. Wells. Hand pump, electric, and solar pump.

7. Municipal or county water lines.

An excellent book on water management is by P.A. Yeomans called *Water for Every Farm: Using the Keyline Plan.* Describing the Keyline Plan is technically beyond the scope of this book. But in summary, on a topographical map, the Keyline is that point in the land where the slope changes from steeper, to longer and flatter. Yeomans states that the significance of Keyline cultivation utilizes two factors:

1. Rainfall on or near a valley, rapidly drains downhill into valleys and off the area. This rapid drainage prevents the ridges from absorbing their fair share of the rainfall and

results in poor soil because the drainage takes with it some of the soil from the ridge and the valleys.

2. Keyline cultivation is literally hundreds or thousands of very small absorbent drains that help prevent rainfall from concentrating in the valley. This helps build and keep soil fertility on the slopes. The land has time to absorb the rain that falls on it.

"The key to Keyline is cultivating parallel to the Keyline". The result is greater water absorption over all the land.

The Keyline plan has been very successfully used in Australia to reclaim and transform dry, barren land into pastures in only a few years. It is a method well worth studying and using. As farmers, we need to know as much as possible about keeping water on our land and maximizing its use.

We believe that two of the most valuable commodities in the future will be uncontaminated top soil and potable water. Bottled water already is more expensive than gasoline. Tons of top soil are lost each year because of needless erosion. As farmers, we need to know as much as possible about keeping water, and soil, on our land.

Chapter 13: What You Don't Know About Electric Fence Can Be Shocking

By far, the most frustrating thing we have had to deal with in the Day Range system is that the electric fence did not produce a shock strong enough to deter an animal. Almost always the fault was in incorrect fence management and improper connections. This caused us to have cows out, foxes in, turkeys wandering, and goats all over the neighborhood. It wasted our time and energy and was a potential danger to the animals as well as humans.

However, once our workers conceptually "got it" not to complete the circle or circuit our fencing problems ceased. Hence, we felt an entire chapter about electric fencing would be useful for you to really use the Day Range system effectively and to avoid some of the problems and frustrations.

Know the Theory - Understand the Flow

The power of electricity comes from the smallest things known to science – electrons. These are tiny particles within an atom. These tiny particles surge through a wire at the speed of light and they always flow from negative to positive.

The charger or energizer is the heart of your electric fence system. Depending on your source of energy, the charger converts main or battery power into a high voltage pulse or "shock" that is carried throughout the length of the fence.

The easiest way to understand this concept is to visualize little men running along the fence, away from the charger, carrying electrons as they go. Most energizers release a pulse (gang of electric men) every second or so. These guys always take the path of least resistance, so any place they can run out of your fence line, they will. Once they get to the end of the fence, they disappear.

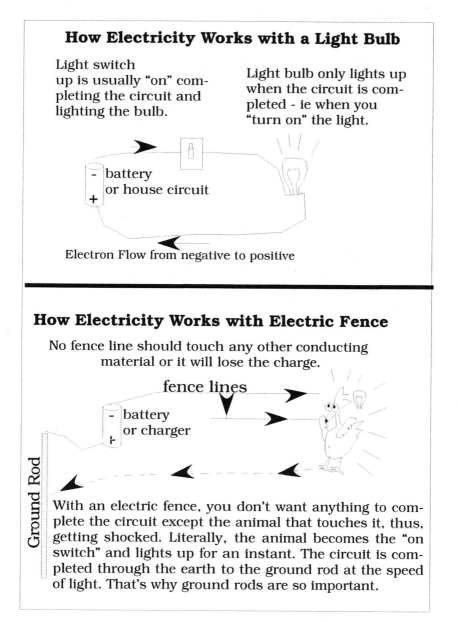

How Electricity Works with a Light Bulb

Light switch up is usually "on" completing the circuit and lighting the bulb.

Light bulb only lights up when the circuit is completed - ie when you "turn on" the light.

battery
or house circuit

Electron Flow from negative to positive

How Electricity Works with Electric Fence

No fence line should touch any other conducting material or it will lose the charge.

fence lines

battery
or charger

Ground Rod

With an electric fence, you don't want anything to complete the circuit except the animal that touches it, thus, getting shocked. Literally, the animal becomes the "on switch" and lights up for an instant. The circuit is completed through the earth to the ground rod at the speed of light. That's why ground rods are so important.

If the fence is expending energy into shocking the grasses that have grown up into it, then it has less energy carrying past that point. This creates a weak spot in the fence from that point on toward to the end of the wire.

Here is a list of ways the little electron men can escape from your system and result in reduced voltage (weaker shocks) in your fence line:

- Vegetation growth touching the live line
- Broken wires, especially in the netting
- Corrosion of wires
- Poor grounding and/or dry soil
- Bad connections
- Poor insulators
- Increasing length of fence
- Connecting fence to itself or to another fence twice. This completes the circuit and you lose your charge
- With poultry netting, hooking the post spike over the bottom line and driving it into the ground under the spike.

All the above are ways the circuit is completed and your charge is drained. So here's what is essential for you to understand. An electric fence only works when the circuit is <u>not</u> completed. The shock itself is the completion of the circuit. The animal becomes the "on switch" and lights up like a light bulb for an instant, until it no longer touches the fence.

Once you understand this basic concept then you will understand not to just tie a knot in the electric tape or twist wire to patch it; you will use a fence connector instead. You will know that if you brace the electric fence to a field wire fence, or a metal fence post, that the fence will short out; you will use an insulated brace instead.

You will not ignore grass and branches touching the fencing, and you will keep the grass under the fence down and remove any tree limbs or branches away from the fence.

Just because the charger is putting out a charge you will not assume the rest of the fence works; you will walk the fence testing it at different points and looking for places where the fence could be losing current and shorting out.

In most cases you will probably be creating a ring or rectangle out of your fencing to enclose an area of pasture or lawn for your poultry to graze. Whatever shape you make, never double the fence back onto itself or attach the two ends of the fence. One end must dead-end. By "doubling back" you can create two pulses traveling through the fence. If these pulses should happen to meet while a person or animal is touching that area of the fence, they would receive a shock greater than what their body could take. In short, it is a matter of safety for you and your flock.

Just because this fence has an electrical pulse running through it doesn't mean it will necessarily shock you. The "shock" is actually created through a completion of the electrical circuit. That's the job of those copper ground rods.

Electric Fence Testers

No one likes to be shocked because it hurts! An electric fence tester is actually a volt meter that lets you determine if there is enough voltage in the fence to shock and control your animals. We keep an electric fence tester in the tool box on the tractor so we always have it handy. The cost of fence testers range from about $15 to over $100. The more expensive testers will tell you the fence voltage, amperage flow and even the direction of the short. We found the six-light tester really hard to read in sunlight. We prefer the digital display models.

We view the testers as cheap insurance because if we don't have a fully charged system, animals can get out, or predators in.

If we need to test the fence but don't have the tester with us then we use a blade of grass. This method only works about half the time. Hold the blade by the fingers of each hand and touch it in the middle of the blade around the hot wire. You will feel a tingle but you won't get the full charge as if you touched it yourself. With the grass method, wearing rubber boots will insulate you from getting a shock and you might not feel any current at all, whereas the fence is working perfectly. If you put one knee on the ground you might get better results.

You're Grounded!

Your charger has two leads; one which connects to the fence, and one which needs to be connected to a grounding rod. By driving a rod deep into moist dirt and connecting it to the charger you create the possibility of a completed circuit.

When an animal touches the fence a current will run from the charger, through the fence, through the animal, down through its feet into the earth and back to the ground rods. The current is carried via moisture in the soil to the grounding rod, and back to the charger and the circuit is complete. All this happens at the speed of light.

The animal is shocked because it becomes a conductor and the current runs through it. When no animals are touching the fence, the electric pulses travel the length of the fence and then stop at the end, to be followed by another pulse only a second later.

Water is a good conductor, which is why fences tend to work better in moist soils. That's why pigs, with their wet noses and pointy little feet in mud really give a squeal when they get shocked. It is harder for electricity to flow in dry soil so in arid soils you will need more grounding rods.

Guidelines for Grounding Your System

- If possible, choose a moist site to place your ground rods. This site can be some distance from your energizer if necessary, but you will need to run an insulated wire to the ground rods.

- Drive three 6 foot galvanized or copper earth rods into the ground. For dry, sandy soils more earth rods might be required. Place the ground rods at least 10 feet apart.

- Connect the earth rods together with one continuous wire. An insulated 12.5 gauge cable is good. Then clamp the cable to the rods.

- For very dry conditions: In dry sandy soil, cold climates, or snow covered or frozen soil, you can use a ground wire running parallel to your hot wire. This ground wire should be grounded every 1,300 feet with 6 foot galvanized steel or copper ground rods. This method of electric fence installation does not depend on favorable soil conditions and moisture. It uses the ground wire on the fence.

Testing Your Ground

If your fence is not giving the shock you expect and you have checked all your connections and have the fence wires insulated, then your problem is probably with the ground system. The main causes of an ineffective grounding are:

- Insufficient ground rods.

- Badly connected wires or wires of different metal types joined together.

- Poor connections to ground rods.

- Rods too close together.

- Rods not long enough or not driven far enough into ground. Depth is important to good grounding because it helps to assure that part of the ground rod is in damp soil.

Connecting Two or More Fences

We usually have our chickens grouped together on one section of our farm, often connecting fences in a series using a solar charger.

Connecting Poultry Netting Fences Together

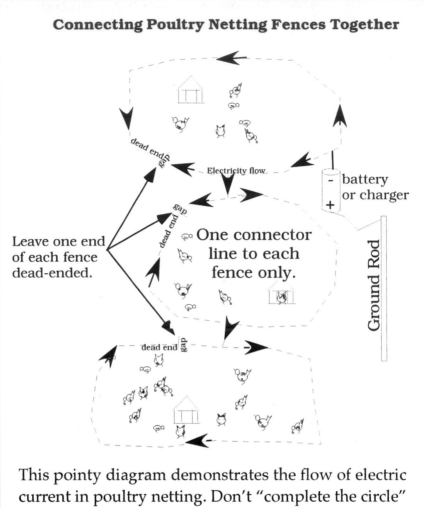

This pointy diagram demonstrates the flow of electric current in poultry netting. Don't "complete the circle" when connecting sections of the fence. Each charged line (fence) needs to dead-end.

Here are a few considerations when connecting fences together:

- Make sure the fence ends are connected (the ends are usually aluminum clips) at only one end.

- Don't overload the battery with too many fences.

- Check that each fence is functioning before linking it to other fences.

- Do make good connections. Use wire clamps, connectors, and good splices. Don't tie knots in the fence line or use tape. Don't simply wrap the wire loosely because this causes corrosion at the splice and reduces the power on the electric fence.

- Do not install ground rods within 50 feet of a utility ground rod, buried telephone line, or buried water line because they might pick up stray voltage.

- Do face a solar electric fence charger toward the south and place where it does not get shaded.

- Do use insulated lead out wires and jumper wires. These should be insulated to 20,000 volt minimum.

- Lightening protection. The only fail-safe method to avoid damage during an electrical storm is to disconnect the energizer from the fence line and from the main power supply. You can install a lightening diverter. Follow the instructions with each kit.

In summary, most problems with electric fencing are usually easily avoided by understanding how an electric fence works. Shorts can usually be detected through careful observation of the fence. One of the first things to do when you lose your charge is to check the connections (charger-to-battery, charger-to ground rods, charger-to-fence, fence-to-fence, etc.).

Then do a "fence walk" around the perimeter to see what may be causing the fence to drain. If possible, we prefer to walk the fence when it is dark so we can see the spark, as well as hear the short.

If all that fails, then perhaps your grounding rods are not being driven deeply enough into the soil or you might need to add more rods, especially in dry soils. The fence that worked fine in May might not work at all in August because the soil is too dry to conduct electricity.

Through keeping the fence free of any interfering objects, and keeping a close eye on the battery, connections and making daily observations you will ensure that your fence is functioning properly and performing its duty. Your life as a Day Range poultry farmer will be easier.

Chapter 14: Processing Poultry

Processing is probably the biggest hurdle to jump for most people. You can do-it-yourself or, if there is a slaughter house nearby, you can pay to have birds processed. Having your birds processed in a USDA inspected plant will cost from two to three dollars per broiler and seven to ten dollars per turkey. The advantage of a USDA plant is that you can legally sell your birds through retail stores and transport them across state lines. This could really open up your market area.

After proper training, one person can easily process 50 broilers or 15 turkeys per day, from start up to clean up. If you have two or more trained people to help, you can increase your processing rates to 100 broilers or 30 turkeys per day. In the beginning, however, before you build up skills and efficiency, it will appear as if you will never be able to process enough birds in a day to make it worthwhile. When we first started out it took us nearly 1/2 hour to process our first broiler. Now we can do 100 per morning, with 4 people.

Processing poultry is easily learned, but it takes a few hundred birds to get good at it. Experienced poultry processors can seem to fly through it, but it will take a novice at least a 1/2 hour each for the first few birds. If possible, visit someone in your area on processing day and have them show you how to do it. Old timers (and equipment manufacturers) will brag about how many they can do in an hour. For example, we can kill over 100 broilers per hour working at a steady, safe pace. However, that's the easy part, and we still need to get them eviscerated, preened, chilled, dried, vacuum bagged, weighed, labeled, and into the cooler. Just remember that all of this takes a lot longer than you think it will when you are just starting out.

Also remember, you can take the best, healthiest bird and ruin it, and all your profit, with poor processing. This happened to us when we contracted out 300 chickens to be processed commercially. They came back with torn skin, broken wings, and looked really awful. We had to discount our price because of the poor presentation.

The phases of processing are:

- Collecting the birds to be processed.
- Killing and bleed out.
- Scalding to loosen the feathers.
- Plucking to remove the feathers.
- Eviscerating to remove the internal organs (offal).
- Preening to get a clean, best looking product.
- Chilling to bring the carcass temperature to 40° F.
- Drip dry to remove excess water before bagging.
- Vacuum bagging and tight sealing.
- Weighing and labeling.
- Into the cooler to age or into the freezer to freeze.

The rest of this chapter focuses on developing your own small-scale, on-farm processing facility. There are essentially four different sizes for processing facilities:

1. The smallest is the one or two person facility, in the driveway or corner of the barn.

2. Next, would be a facility for 4 to 10 workers, probably with an indoor, screened space.

3. Larger still is a commercial facility with 10 or more workers, which is USDA or State inspected.

4. Finally, the fourth model is a Mobile Processing Unit (MPU) that travels from farm to farm, and usually relies on farm members for labor. This system is still being studied and there are legal ramifications that vary by state.

Let's examine the advantages and disadvantages of each.

1. Smallest – for One or Two People

In this size facility, the grower does the processing either alone, or with the help of a spouse or partner. This is typically how we all get started. The equipment needs for a 1 or 2 person operation are small: a shackle for killing, a tub of hot water for scalding, a one-bird drum plucker for de-feathering, another shackle for eviscerating, and a chill tank. All of this can cost anywhere from nothing to less than $1,000, however, it is exceedingly slow.

Here's an average time estimate for an efficient one-person operation using basic one-person equipment for 60 broilers in a ten hour day:

Average Time/Bird and Task

1 minute to catch and crate (night before). 60 minutes total.

1 minute to get water hot and prepare the facility: sharpen knives, set up stations, and deliver the crated birds to the facility. 60 minutes total.

1 minute to pull a bird from crate, put in shackle, cut and let bleed. Use two or more killing shackles so that while one bird is bleeding out you can be killing another, or move one to scalding, or doing any of the other tasks needed. 60 minutes total.

45 seconds to scald, with water temperature at 142°F. Either dunk the bird up and down by hand, or swish around with a paddle.

15 to 45 seconds to pluck (table top or floor model drum picker).

15 seconds to remove feet and head.

1 minute to eviscerate, preen, and place in chill tank with ice water at near 32°F.

1 minute to remove chilled bird from chill tank, drain, and hang to dry.

1 minute to bag and vacuum seal.

1 minute to weigh and label, place in reach in cooler, refrigerator or freezer.

1 minute to clean facility. 60 minutes total.

Average 10 minutes total/bird, or 10 hours for 60 birds.

The total is bout 10 minutes per bird. This includes 1 hour the evening before to catch and crate 60 birds, then eight to ten hours on processing day to get 60 birds into your cooler. It is a full day's work. Processing a turkey takes at least three times longer than processing a chicken.

In the beginning, before we switched to shackles, we used killing cones to bleed the birds. We didn't like the cones because the birds were hard to get in, often requiring two hands just to get the head and neck through the bottom opening. And, sometimes when the throat was slit, the bird would panic and back itself out of the cone, falling to the floor and thrashing around spewing blood all over the place. Not fun or humane. We also found that with cones the birds didn't bleed out completely. This was especially true for blood pooling in their wings, which were held tightly against their body by the cone shape. We no longer use cones.

Now we use shackles. It is much easier and quicker to get the birds legs into the slots. The shackle holds them securely while they bleed out completely.

There are several ways to get the birds squeaky clean before you bag them for your customers. When we are ready to start eviscerating each bird, we check its arm and leg pits quickly for tiny feathers and over all for any pinfeathers or blemishes to be removed. Then, after they have chilled in the chill tank, we hang them back on the evisceration shackles and double-check them for any pinfeathers or bits of innards that still remain. While they are hanging on the shackle they drip dry, then we slip a plastic bag over them and hand them over to the vacuum bagger.

Before we started using the shackles to hang the birds for drip drying, we used a drain frame made from a board with 1-foot long 1-inch diameter PVC pipes mounted every 8-inches. We would sit the carcass on the drying pegs and let the water drain off while we were doing the final preening and then slip a bag over the bird.

If you don't get the excess water out of the carcass, your final package will have a bloody pool of water at the bottom. That is unsightly when you present the birds to your customers. Even worse, bags can get small holes in them and the water leaks out in your refrigerator and makes a disgusting mess.

Before we could afford a commercial Tipper Tie vacuum bagger, we used a shop-vac to get a relatively decent vacuumed package. To maintain the vacuum, we held the bag tightly closed with our hand as we removed the nozzle from the plastic bag. We sealed the bag with hog-ring pliers. The hog rings give a semi-good seal, but this process is slow. Now we use a Tipper Tie vacuum bagger that does all these things and moves swiftly and securely. See USDA processing facility section for complete description of Tipper Tie Vacuum bagger and clipper head.

To get the air out of the bag, some people submerge the bird in a bucket of water. This is messy, time consuming and ineffective. Another problem is that your bag is wet and tape or labels won't stick on wet bags. Some growers simply squeeze the air out by hand before they put on the twist tie. In my estimation this results in a poorly finished product and is not acceptable.

If you don't get most of the air out of the bag the meat frosts and gets freezer burn quickly. This dramatically shortens the shelf-life of your product, as well as making it unappetizing.

We recommend that the very least you should do is to let the bird drip dry for a minute or two. Put it in a good quality freezer bag. Use a vacuum to pull out the air and excess water. Seal the bag with a metal clip and hog-ring pliers. Bags and hog ring pliers are available from equipment dealers listed in the resource section.

I suggest that you have your customers pick up chickens on the same day you process. Have them come later in the afternoon when you are totally finished, including cleanup. Each customer visit can take from 5 minutes to 10 minutes and can be distracting if you are still processing when they arrive. Answering simple questions, and starting and stopping can quickly take up lots of time. Even after you get reasonably proficient at all phases of your processing system, you will have all you can do to get that many birds processed at one time.

2. Small – a Four-person Processing Facility

Once you have outgrown the "one bird at a time" processing facility you will want to look into faster equipment that is more efficient. We have a 4-person processing facility added onto our barn, and with a full crew we can process 100 broilers in a morning.

This fits nicely with other farm chores. Whoever is doing morning chores lights the fire under the scalder to get the water hot, and brings the loaded crates of broilers from the field. This is done in the same trip when they come back from feeding and tending the field flocks.

We use plastic poultry crates for capturing the birds in the field. We find it is much easier to catch the birds at night and put them in crates. They are virtually night blind and won't wander very far. We take our time and try to catch them gently and quietly, usually picking up 2 to 4 birds at time by their legs. We leave crates in the field overnight so the manure will go on the ground for fertilizer. We collect the crates in the morning and deliver them to the killing area.

By the time morning chores are done, the other three members of the processing crew meet in the barn. While one or two are filling the chill tanks and getting the knives sharp, the other two unload the broilers from the tractor and into the kill room.

The area needs to be large enough for a 4-person crew to work efficiently without getting in each other's way. We have two separate rooms; one for holding the birds, killing, scalding and plucking, and another room for eviscerating, chilling, and bagging.

One person does the catching, killing, scalding, plucking, and removing feet and heads. They then hang the carcass on shackles in the evisceration room.

Three other people remove any leftover feathers, then eviscerate and place the birds in a chill tank. It takes up to two hours for the birds to chill to 40° F. Then they are removed from the ice water and are hung to drip. A quality control person checks them for blemishes and pin feathers. A second person bags the birds, vacuums out the air, and seals each bag. Then the birds are weighed, labeled, and put into refrigerator.

The Four Person Recommended Equipment:

Poultry crates for catching and holding broilers: Turkeys are simply herded into a temporary pen near the processing room and held until needed.

Killing shackles: We prefer shackles instead of killing cones because it is much easier and is twice as fast to get the birds in the shackles, and the birds can't back out of them. We use

16 shackles mounted on a 36-inch diameter hub. You will also need a sharp knife for slicing the carotid artery for successful bleed out. Don't cut the windpipe too, because the bird will suck blood back into its lungs and die prematurely before being thoroughly bled out.

Electric stunning knife: Most growers don't use one because they are quite expensive, about $1,400. We use one because it makes the kill task much easier, faster, and more humane.

Once you touch the bird's head with the stun knife it is unconscious and doesn't squawk and flail much. This makes the birds easier and safer to handle and it saves time for the kill portion.

Once stunned, bird's relax their feathers which makes them easier to remove. The result is a more complete pluck. That saves time in quality control. The relaxed bird also gives a better bleed out for a cleaner carcass. That gives a higher quality, more appealing product.

We believe the stun knife is one of the best investments a commercial grower can make. By the time you do 10,000 birds the cost will average out to only about 14 cents per bird, not to mention the savings in time for your crew and the superior product that results.

Scalding tank: One large enough to have a mass of water that won't cool prematurely when birds are dunked. You can use something as simple as a 15-gallon galvanized wash tub on an outdoor cooker. We use a 8-bird Brower rotary scalder that has a revolving shelf to dunk the birds in and out of the water. A built-in gas furnace heats the water.

Plucker: The drum pickers are least expensive, but more difficult to learn how to use effectively and efficiently. The can only do one bird at a time. We use a 4-bird Brower batch plucker that has a built-in water spray and timer.

Evisceration shackles: These are hung over an offal barrel. Some growers use table top evisceration, but the shackles are much faster and a whole lot more sanitary.

Water supply: for rinsing and cooling birds, and for washing down your facility. A pressure washer (about $130) will make clean up a lot faster and better. The pressure washer has a built in soap dispenser for soaping down all surfaces, and then pressure rinses everything clean.

Chill tank and ice: We use a 100-gallon polyethylene stock tank, and buy ice from the local convenience store.

Vacuum bagger and clipper: Some growers just take the bird out of the chill tank and put it in a plastic bag and close it with a twist tie. This is not very attractive and the bird will be subject to freezer burn. It is much better to re-hang the bird on the shackles to let it drip dry, then bag it, vacuum out the air, and close the bag with a metal clip. You can use something as simple as a wet/dry vacuum and hog ring pliers. We use a Tipper Tie semi-automatic vacuum and clipper.

Refrigeration: You need to cool the birds in a refrigerator to less than 40°F, and then in a freezer to hold them until you sell them. Letting the birds "age" for a couple days before freezing helps make the meat more tender. We cool them to 38°F and hold them there for several days before putting them in the freezer. We have a reach-in Hobart commercial refrigerator that will hold 100 broilers. Only stack the birds one layer high until they are frozen solid. Using a second layer will squash the birds beneath, giving you a very unattractive package.

Clothing: All personnel will need a hair net, a plastic work apron, and rubber boots. Some workers prefer to wear latex gloves. This is especially important during the final preening and bagging operation, so that dressed poultry doesn't get contaminated before being placed in the bag.

Knives and sharpening steel: It is very important to have a sharp knife to make the cuts. Dull knives take too long and can lead to accidents.

Buckets and containers: You will need food grade plastic buckets for offal and for giblets. Put the giblets in the carcass in zip lock bags.

Safe food labels: We tape the food labels along side our farm label. We feel it gives a professional look, as well as possibly cover us legally if someone does not handle the meat safely after it leaves our farm. The labels themselves are only 2"x1" and the type is so small one needs a magnifying glass to read it; but it still has good information and is worth the effort to add it. You can order these labels from Brower, and other food processing supply vendors.

Packing table: A well arranged packing table is a timesaver. It needs easy access to all the scales, bags, ties, vacuum, knives, and totes to carry the packaged birds to the cooler.

Clean up: You will need good hoses and nozzles for rinsing birds and washing down rooms. Leaving birds in the chill tank for a while after evisceration will help scratches and blemishes leach out.

Gloves and hand protection: Use rubber or heavy duty latex gloves for killing, scalding, and plucking. Some folks use them for eviscerating as well, but they really are harder to use than bare hands, unless you have open cuts, in which case you must use latex gloves. Even

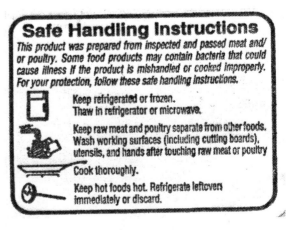

Safe Handling Instructions

This product was prepared from inspected and passed meat and/or poultry. Some food products may contain bacteria that could cause illness if the product is mishandled or cooked improperly. For your protection, follow these safe handling instructions.

Keep refrigerated or frozen. Thaw in refrigerator or microwave.

Keep raw meat and poultry separate from other foods. Wash working surfaces (including cutting boards), utensils, and hands after touching raw meat or poultry

Cook thoroughly.

Keep hot foods hot. Refrigerate leftovers immediately or discard.

consider using a protective kevlar glove for left hand during certain operations.

Some of you have asked why we advocate using shackles instead of table top evisceration. The four biggest reasons are:

1. The operator stands upright, working at eye level, eliminating backache from leaning over a table, and using gravity to let the guts fall straight down into a gut bucket or trough.

2. It's faster with a shackle because you can use both hands to do all the cuts, eviscerate, preen, and wash down with a hose attachment, all in just a few seconds per bird

3. It's cleaner with a shackle because the bird is suspended and any blood or manure falls down to the trough. You can also see to wash down and preen the bird because it is right at eye level.

4. With the shackles mounted on a rod or conveyer, your system can speed up by having workers specialize on each task. For example, I can do the cuts, then push the bird to the second person. They scoop out the guts and send the bird to the third person who takes out the lungs and windpipe and passes the bird to the fourth person. The fourth person preens, pulls the head off, and cuts the neck off. They then drop the broiler into the chill tank. The last person in line moves the shackle hanger to a second rod and scoots it back to the beginning where the person who does the killing and plucking can hang the next bird. We've just bought a used rotary line that will automatically carry the bird in front of all line workers and then return the empty shackle to be refilled.

The hardest part of getting folks to switch to using shackles is that it is all but impossible to describe to someone how to use the shackles unless you are there to show them.

Detailed Description of Processing

Below is a rather long and detailed description of how we process birds after they have been through the plucker. This method is very similar to what is used in USDA processing facilities. The difference is that in large plants, each step is done by a different person or machine. In this small system, you might be doing all the steps yourself. Even though you can't see his face (like Wilson on "Tool Time"), Andy is the star processor in all these photos — he's the one without wings.

1. Cut off hocks. Once the bird is plucked, remove the feet by using pruning shears or a sharp knife to cut through the hock joint. The hock joint has two ball-shaped bones connected by skin and sinew. Bend the leg slightly to locate the ball joint and cut through the skin and tendons for a clean separation. Leave the head intact.

2. Hold the bird in front of you with one hand on each thigh, breast facing you, and head down.

3. Place the hock joints in the shackle so the hock joint is "behind" the shackle, the leg is in "front" of the shackle, and the breast is still facing you.

4. Bend the bird upward and place the head in the shackle leg on the left. If the bird won't bend then you have it in the shackle wrong, with the hock joint in front and the leg behind the shackle

5. Use a sharp knife to slide under the skin and make a slit in the neck skin going up from between the shoulder blades to the back of the head.

6. Tilt the bird top away and the bottom toward you so that the oil gland located at the top of the tail, is at eye level facing you. Cut the oil gland out. Some people prefer to cut the entire tail off, but I just do the oil gland.

7. Let the bird swivel on the shackle hanger so the vent is facing you at eye level. Make a "U" cut around the vent, being VERY careful not to insert the knife too far and puncture a gut.

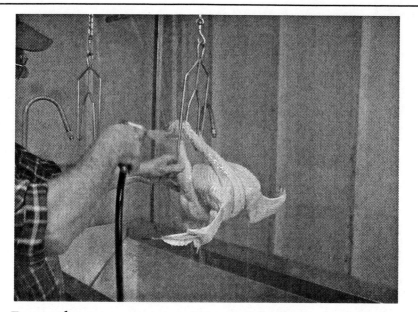

Preen the carcass removing any feathers the plucker missed and rinse well. Notice how the bird is suspended by the shackles.

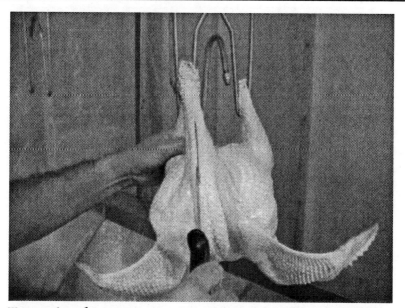

To make the neck cut, start between the shoulder blades and cut upward. Notice how Andy keeps his left thumb on the thigh and not on the neck to hold the bird steady. It is well away from the blade should he slip.

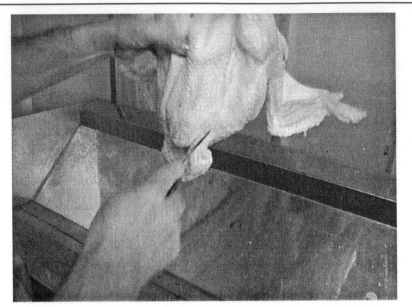

The U-shaped vent cut. Be careful not to puncture the intestines. Andy keeps his left hand well away from the blade so any slip won't cut him.

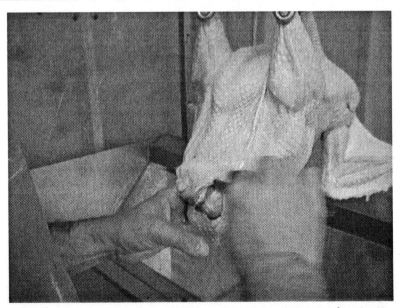

Insert the knife above your finger and cut the belly pad out of the bird. This makes a big opening for your hand to clean out the eviscera.

8. Assuming you are right-handed and are holding the knife in your RIGHT hand. Use the index finger of your LEFT hand to enter the "U" cut just above the vent, and separate the belly skin from the entrails, leaving your finger in place separating the vent and gut from the belly pad. Insert the back of the knife above your left finger. KOCH (1-800-777-5624) sells a special vent knife that has a sloped blade back so you don't inadvertently puncture your finger.

9. Once the knife is in position, parallel with your finger and the blade back against your finger, cut the belly pad out of the bird, being VERY careful not to cut a gut. Cut the belly pad all the way to the end of the keel bone. This makes a nice big opening for your hand.

10. Now put the knife down — you are through with it until the next bird.

11. Some people don't cut out the belly pad but instead they put a slit in it so they can slip the hock joints through and lock the legs down. I don't do that because it looks unsightly and because it makes it impossible to check for pin feathers in the leg pits.

12. With the back of the neck facing you, work your fingers and thumbs along the neck to break away the fascia, loosening the windpipe, crop, and jugular vein. Separate everything from the neck AND the neck skin. Work your thumb down inside the wishbone area to loosen the crop all the way inside the cavity. This whole area has to be CLEAN, and many people don't take the time to do it, but you will have a clean, wholesome product if you do it right.

13. Unhook the head from the shackle and either pull the head off, or cut with a knife, or use pruning shears. Throw the head away or save for dog food. Double check the neck to make sure all fascia, windpipe, and veins and crop are loose. If they are entirely loose they will pull easily out the rear when you remove the guts.

14. Swivel the bird to put the vent facing you. Have your left hand supporting the bird by gripping around the barrel at the wing pits. Pull gently on the vent to extricate some of

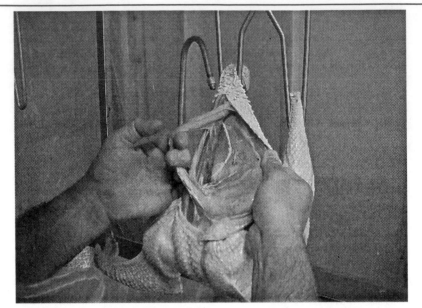

Work your fingers and thumb along the neck to separate the fascia, crop and windpipe. If done right, the crop and windpipe will come out with the gizzard and internal organs when you pull them from the rear of the carcass.

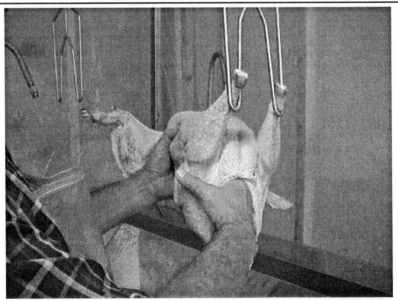

Put your hand inside the carcass. With your fingers loosen the eviscera on both sides so that when you pull the gizzard almost all of the other internal organs will also come out.

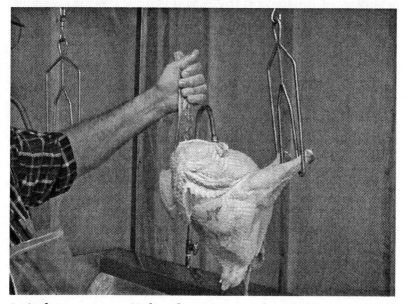

Lift the carcass off the shackles by the neck.

the entrails to make room for your hand inside the cavity. Careful, don't break a gut. Enter your right hand into the cavity, and work from left to right to loosen the viscera from the rib cage and back bone.

15. Work your hand forward until your fingers have loosened all the viscera, and you can feel the heart and gizzard. Cup your two fingers around the gizzard, and ease the whole mass of entrails out of the bird. Take your time and coax them out rather than prying or jerking that might break a gut. If you did everything right, all of the guts, the gizzard, the windpipe, and crop and jugular vein will come out as one mass.

16. Put your hand back inside the carcass and feel the ribs, then use your finger tip down inside the rib cage to loosen the lung first on the right side and then on the left side and pull the lungs out. Sometimes a piece of windpipe is still attached to the lungs.

17. Swivel the bird so you can slip your right index finger inside the wishbone and peel away any remaining crop or windpipe or jugular or any fascia that is still attached. If it

is yellow its fat, leave it there. If it is flesh colored it's viscera and needs to come out.

18. With a hose nozzle, fill the cavity of the bird and rinse out any remaining blood or body bits. Recheck the wishbone area for last little bits.

19. Preen the bird, looking inside the cavity for stray bits, pick off any pin feathers, and trim off any scabs or blemishes. Rinse the bird thoroughly so you don't put a bloody or dirty bird in your chill tank.

20. With your left hand, grab the neck and lift the bird out of the shackles. Use pruning shears to sever the neck at the shoulders letting the bird fall into the chill tank, and pitch the neck into an ice bucket.

21. Return the empty shackle for the next bird.

22. Put enough ice in your chill tank to get the birds down to 40°F. This usually takes about an hour or two. A 100-gallon stock tank will hold 100 broilers averaging 4 pounds dressed weight, and you will need about 400 pounds of ice in really hot weather. Having an air conditioner in your evisceration room helps keep things cool, especially your people.

23. When the birds are properly chilled and you are ready to bag them, rehang the birds on the shackles. After they are chilled you will be able to see blemishes and pin feathers more easily. Preen the birds a final time. Let them drip dry to remove as much of the chill water as possible.

24. Use good freezer-proof bags (BROWER 800-553-1791) and bag the bird with the neck skin folded back across the neck stub, and the bag molded down around the bird. Hold the legs down against the body, insert a narrow vacuum nozzle into the bag and suck out all the air until the bag is collapsed tight against the carcass and the legs are tight down.

25. You will have to insert the vacuum nozzle before you turn on the vacuum.

26. Twist the neck of the bag tightly, and staple it closed with hog wringer pliers and 3/8-inch rings. The hog wringer

pliers are available for $19.95 and the hog wrings are $1.50 per hundred from The Sausage Maker (1-888-490-8525) in Buffalo, New York.

27. Chill the broilers for at least 24 hours in the reach-in cooler, then put them in the freezer. Only stack them 1 layer deep until they are frozen so they don't mash each other out of shape. Once they are frozen you can pile them several layers deep. The 24 hours chilling time before freezing helps tenderize the meat, and improves flavor.

Well folks, if you have made it this far you are probably pretty discouraged at the idea of ever doing processing or using shackles. However, after you've processed your first hundred birds this way we can guarantee you will never want to go back to table top evisceration. The shackles are faster, cleaner, and easier on your posture and back.

3. USDA Inspected Facility

Based on preliminary investigation, it is possible to build a small-scale USDA-inspected facility for less than $40,000, not including cost of land. This does include the cost of a small concrete building with holding room, kill room, evisceration and packing room, walk-in cooler and walk-in freezer, bathroom and office. It does not cover the cost of an independent septic system for the bathroom, a second drain field for waste water, and an independent well for water supply. This could add several thousand dollars to the cost. It will be well worth your while to get a copy of the USDA regulations and work with one of the inspectors before you begin building your plant.

Listed below are the basic components you will need to get UDSA inspection.

- An office area for the inspector. This can be as simple as a desk and phone in the corner of the room.

- Walk-in cooler is accessed directly from the evisceration room, and the walk in freezer is accessed through the cooler.

- Washable walls and floors, to meet USDA approval. Floor drains at least 4″ in diameter, and preferably 4″ gutter type drain for fast wash down.

- Bathroom with shower for staff, and possibly a second bathroom for the inspector, depending on how large your plant is.

- Storage closet for tools and supplies.

- Drains to septic field for wash water, and a compost area for eviscerate.

- Holding pen at kill end.

- Killing shackles and blood vat with drain, scalder with drain, plucker with drain area, evisceration line for three or four people with drains and water lines. Screen doors to keep out flies and wasps/yellow jackets out. Add an air conditioner through the wall to keep temperature at a comfortable level, and later in the year a heater.

4. Portable Processing Unit

A Mobile Processing Unit (MPU) travels from farm to farm, and usually relies on farm members for labor. This system is still being studied and there are legal ramifications that vary by state. Here are some thoughts on portable or stationary facilities.

To get a processing plant approved there are at least three hurdles:

One is USDA approval. In Virginia, at least, this is easier because of the Talmedge Atkins Act, which enables state inspectors to do USDA inspections. It remains unclear whether the state or USDA will actually approve a portable facility in the state of Virginia, or any other state for that matter.

The second hurdle is the Board of Health in the county. They will want to approve the facility—especially if it includes a commercial kitchen—as well as the septic system.

The third is the zoning administration of the county. According to our local county zoning administrator, slaughterhouses are prohibited in agriculture zones because agricultural processing is different from agricultural production. Processing is only allowed by special permit in industrial zones (which are quite expensive). This is probably true throughout most of the United States.

We think the obvious reason for this is everyone's fear that some huge conglomerate will want to put in a 10,000-cow per day slaughterhouse, or something of that nature. Our hope here in our county is that the county commissioners will amend the zoning restrictions to allow small-scale (say up to 50,000 birds per year) poultry processing. We are currently attempting to pass this zoning amendment. It is also not clear at this time whether a PORTABLE processing facility—one that only visits a farm one or two days per month— would come under county zoning restrictions.

In order to obtain USDA approval of a portable facility we will need to begin with a basic vehicle and remodel it with washable walls and floors, floor drains, a curtain wall between kill room and evisceration room, chill tanks, and clean up facility. All new equipment will probably cost about $15,000, plus the cost of buying and remodeling the base vehicle at about $10,000, for a total cost of $25,000. It appears unlikely to us, however, that the USDA will approve such a portable facility. That then begs the question, "do we really want or need USDA approval?"

The vehicle requirement for a mobile processing unit would be big, at least the size of a school bus. However, a school bus wouldn't work because of the low ceiling (74"). By the time we built up the floor to install drains, it would be too cramped.

A truck similar to a Ryder or U-Haul would be okay, (about 8' by 24') but doesn't give a whole lot of room. It might be possible to equip a trailer with the kill room equipment (i.e., kill cabinet, scalder-dunker, plucker), and then equip the truck box with evisceration table, chill tanks, and packing table.

We haven't yet made estimates on the cost of a water supply, clean up procedures, composting offal, managing waste water from processing, clean-up and so forth.

This also does not include the additional cost at each farm for other USDA requirements. These include an office for the inspector, a bathroom for the inspector and processing staff, a septic system that will be required for the bathroom, and a second drain field for the waste water from the processing and clean up of the facility. This also doesn't include the cost of a walk in cooler or freezer.

I envision about 8 or 10 farmers within a 2 to 4 hour radius making up the producers' cooperative. And I envision at least a 4-person trained crew to accompany the processing facility. Does it make sense to have 4-person processing crew traveling for two to four hours a day to each farm and then return that evening? Or, does it make more sense for each farmer or assistant farmer to travel for 2 to 4 hours to bring chickens to a central facility?

Wouldn't it be easier to hire and retain trained people in a central location? And, wouldn't it be easier to work with the USDA or state inspector at one central facility? Especially with the new HACCP regulations?

Ultimately, we believe it will be cheaper and more efficient to build one central facility rather than a mobile processor, including the cost of the portable processor, the office, bathroom, septic, walk in cooler and walk in freezer on each farm. A complete, turn-key, USDA-approved central facility might cost only $50,000 to $100,000. Whereas, a portable unit could cost as much as $40,000 plus an additional $10,000 for each farm, for a total of maybe $50,000.

We are really leaning towards a central facility as being much more efficient and productive, and ultimately easier to build and to get USDA approved.

Then there is the problem of what to do with the processed birds at each farm if we use a mobile processor. Do we have the delivery truck go around to each farm and pick up the birds then deliver them to market? Or, wouldn't it be easier to have all the birds in the cooler at the central facility and have the delivery truck load out an optimal load weekly for delivery?

There are many unanswered questions to mobile processing units. Steve Muntz, at the Heifer Project in Kentucky, is one person who is trying to answer these questions and get an MPU approved. He is working with funding from USDA/SARE. Steve has build and equipped a prototype MPU and is working his way through the process of getting it approved.

Processing Turkeys

Processing turkeys involves the same steps as with chickens, except that they are a lot bigger and harder to work with.

To transport the turkeys to our processing facility, we either herd them, or use a small S-10 pickup truck with cab. We just load the turkeys in the back and back it up to the killing area. We off-load them as we are ready to kill them.

I usually say a prayer before killing each bird. Just a short prayer for these magnificent birds.

Use 5-gallon plastic buckets with a 4" hole cut in the bottom for killing cones. This will work on birds weighing from 12 to 30 pounds. For smaller turkeys, you can use broiler shackles. For the huge turkeys, tie their legs to a sawhorse or tree branch and let them hang as they bleed out. We use tree pruning, lopping, shears to cut the heads, neck, and feet off the turkeys.

Chapter 15: Predator Prevention & Control

You need to pay special attention to predator prevention. Some form of electric barrier to keep the predators out of the flock is critical. Over the course of a decade of growing poultry on pasture we have had death losses caused by rats, foxes, coyotes, dogs, possums, raccoons, hawks, and owls. Other predators that we have not encountered, but that might cause you problems are bears, badgers, minks, weasels, and feral cats.

The only proven way to stop all of the terrestrial predators is with an appropriate electric poultry netting, or by using expensive welded wire mesh fencing and fence skirts to keep predators from tunneling under them. Even then, some predators such as raccoons will climb over a welded wire or mesh fence, and foxes, coyotes, and dogs will jump a fence to get at the poultry. If you use a mesh or wire fence, it is important to also put an electric hot wire outside the perimeter fence to keep predators away.

Under normal circumstances, predators such as raccoons and possums will find a way to push under the fence to get at the poultry. Larger predators such as foxes and coyotes will approach the fenced area and place their paws at the top so they can look over the fence to judge its height. Then a clever fox will back off a few paces and get a running jump to go over the fence. The best way to stop them is to make sure they get shocked either when they approach the fence, or when they put their paws on it to test the height. Once they are shocked, they will generally run away and not come back for some time.

Various other methods of predator prevention will work, at least for a little while. For that matter, you could even put your poultry out unprotected in an open field for a while before they were attacked. Sooner or later though, predators will try to get your poultry.

Some growers report that they are relying on a single hot wire about 6 inches above the ground, in a circle around the poultry. This is not a good plan, since a fox, coyote or dog will simply jump over the single wire, and we've even seen them jump between hot wires on a high tensile fence without getting shocked.

We keep predator numbers low at our farm by hunting and trapping them. We use a large Havahart trap for catching raccoons and possums, and a small Havahart trap for catching rats in the barns and brooders. Just place the trap at a place where you see the predators are scooting under the fence.

Last year we caught 6 possums, 1 raccoon, and 4 chickens at the same path under the fence to the layer flock. Havahart traps are wire box traps that do not hurt the captured animal, and will contain it long enough for you to relocate the critter or dispose of it. They are available from hardware, feed stores or mail-order catalogs.

Our cats are highly skilled mousers, ratters and molers and keep these populations under control. Birds are another matter. There isn't any part of "NO" our cats don't understand. We really fuss at them if they catch any kind of bird. They know not to bother baby chicks, and will actually protect them.

Reach out and touch someone. This is Possum, one of our specially trained poultry protectors and the designated night-shift-rodent-master.

We saw General, our Maine Coon cat, save some baby guinea from our neighbors cat who was stalking them. General got between the guinea keets

This is our gang-of-three farm defenders and fox-chaser-offers. They are full-blooded "hybrids" with registration adoption papers from the Humane Society. On the left is Doodle, on the right is Beamer and in the middle is Lucky Woody Womac.

and the other cat and challenged him. With all the fuss, the dogs joined in and the team of them ran the cat off the property. The baby keets were very safe — amazing but true.

We also rely on our 3 terrier-cross dogs to keep burrowing predators away from the farm. These are full blooded hybrid dogs from the Humane Society–complete with papers.

The dogs are extensions of our senses and hear things we don't. They are very good at warning us if foxes or stray local dogs are on our land.

We do not use leg hold traps because we believe they are an inhumane way of predator control. We also don't want any of our 4 cats or 3 dogs to get caught. We don't use poisons for the same reason.

Andy's Hunting Advice

By far, we have lost more birds to foxes than any other predator. We support wildlife, but when a fox hunts it will kill lots of chickens, just leaving the carcasses behind. Like people, foxes hunt for the play and sport of it, without any intention of using all the kill for food. Their sport is our pure profit loss.

And that's why I hunt the foxes.

I use either a 12-gauge shotgun loaded with "buckshot" or a .22-gauge Magnum rifle with a scope. Sometimes I can get a clear shot in the daytime as the foxes cross our field from one wood lot to the other. Usually though, I have better luck calling the foxes with special tapes that mimic a wounded rabbit scream. The fox will come in close to investigate, then I shine a spotlight to temporarily blind the fox.

The flood light gives me enough light to see through the rifle scope to line the sights on the fox. Usually a single .22 magnum bullet will not drop a fox in its tracks. It will run a few yards and then fall over. I don't use a larger gauge rifle because I don't want the noise waking up our neighbors, and I don't want to risk a bullet leaving our property and hitting somewhere else.

I also have a .22 automatic rifle equipped with open sights and a 5-battery Mag light taped to the underneath of the barrel. If the dogs tell me there are chickens in distress in the middle of the night I can use the mag light to locate the trouble, then temporarily blind the fox while I get off a shot or two.

At 1 a.m. one summer night I stood on the balcony of our second floor bedroom and shot a fox about 50 feet from the house. The fox weighed 9 pounds and had just killed an 18 pound Sweet Grass turkey hen. That fox had the most beautiful, thick, shiny coat I have ever seen on a wild animal. He was a real testimony to all our organic chicken he had eaten over the course of his life.

As far as rifle selection, if I had it to do over again, I would buy an over-and-under rifle and shotgun combination. That way I can use the shotgun for close-in shots, and the rifle for longer distances. I would equip the rifle/shotgun with a 4x power scope with a wide angle lens that gathers light for use at night. I'd also install a thumb operated spotlight mounted on top of the scope.

The spotlight has a red lens for scanning terrain without spooking varmints. Once the varmint is targeted just slip the red lens off to spotlight the varmint and temporarily blind it while getting off the first shot with the rifle, and a follow-up shot with the shotgun. There are also night vision scopes and laser sighting devices that extend the usefulness of your firearm.

The hunting rifles, spot lights, and varmint calls can be purchased from your local sporting goods dealer. Always remember, guns are dangerous, so handle them carefully, and practice as much as you can so you will be able to handle the gun safely and efficiently. While I was in the Navy, I was lucky enough to compete in the NATO Goodwill Games and won the marksmanship championship, so I'm very comfortable with rifles.

Airborne predators are a special challenge. You can protect your flocks from owls, which only hunt at night, by locking your poultry inside the shelter from sundown until chore time the next morning. Hawks will take a few of your birds, more in the spring when some hawks are migrating.

The larger your birds before you put them on pasture the better. Some people have reported luck in trapping hawks and owls. They place a platform on a pole at the edge of the chicken yard, and install baited Havahart traps on the platform. Then they relocate the raptor out of the area. As far as we know, trapping or shooting any migrating bird is unlawful.

The electric poultry netting offers the best hope of keeping your poultry safe. It is expensive, about $1 per lineal foot plus the cost of the charger. And it requires a certain amount of time to install and keep electrified. However, year in and year out, the electric poultry netting has cut our losses by predators to almost zero.

We have been using electric poultry netting from several sources, and find there really isn't much difference in cost or performance. However, our favorite is the British Electranet sold by Kencove (see resource guide). For a while, we were using their model NP-7, but found that the smaller broiler chicks could slip through the 7-inch stay spacing too easily, although it works fine for layers and turkeys.

For the smaller birds we use poultry netting with 3.5 inches between stays, and find it to contain the smaller chickens much better. It is important to make sure the fence is really "hot" when you release poultry into the paddock. Their first experience with the electric poultry netting should be "shocking"! Once they understand that the shock comes from the fence, they will steer clear of it.

Other people have reported having good luck with large guard dogs. They fence in the entire area to hold the dog in, then let the dog free range in amongst the various batches of poultry. Any predator that comes through the perimeter fence alerts the guard dog, who will deter or kill the predator.

Turkey Breeds versus Varieties

Views of the Authors and Publisher

In *Day Range* we use turkey "breeds" rather than "varieties". By doing this, we acknowledge that we are going against convention and textbook terminology.

However, we strongly feel the use of "varieties" does turkeys a great disservice, and that to use "breeds" is technically correct. According to a definition taken from *A Conservation Breeding Handbook,* published by the American Livestock Breeds Conservancy:

What is a Breed?

"Breed" is an important concept, and its definition must be appreciated before purebred breeding or conservation has much meaning.

The best definition of a breed is *"a group of animals selected to have a uniform appearance that distinguish them from other groups of animals within the same species. When mated together, members of a breed consistently reproduce this same type."*

If that's good enough for chickens, pigs, goats, cows and sheep, then surely it's good enough for turkeys!

Today, here and now, the preservation of the heritage turkey breeds is at a critical point. All but the two commercial industry breeds are in danger of becoming extinct — that means forever gone! We are referring to breeds like the Black Spanish, Narragansett, Royal Palm, Bronze, Slate, Buff, White Holland, and Sweet Grass. These are all "groups of animals within the same species that consistently reproduce the same type". They are turkey breeds.

To us, it seems much more compelling for folks to act and save a vanishing breed rather than a variety. "Variety" seems more like something you would pick up at the corner discount dollar store—something not valuable. We apologize in advance if using turkey breeds offends any individual or organization, for that is not our intent. Our intent is to help preserve the turkey genetic pool and the viability of different turkey breeds for everyone's future — forever.

Frequently Asked Questions (FAQ)

About Pasture Poultry

We encourage readers to copy any part of this that might be helpful to them in educating their customers.

Q: Why raise poultry on pasture?

Raising poultry on pasture appeals to consumer concerns about humane treatment, healthy living conditions, clean food and good flavor. The Day Range production model offers the grower an opportunity to earn a sustainable farm income.

Public perception in some areas is that antibiotics, growth hormones, and meat by-products are routinely fed to commercial poultry to increase profits for large integrator corporations. Additionally, some people feel that concentrations of poultry houses in any given area can lead to environmental degradation and real estate property devaluation. Some people also feel that commercial poultry are treated inhumanely, and that some diseases and illnesses are caused by overcrowding in huge barns without proper ventilation, long-distance transporting in all kinds of weather, and unsanitary commercial processing facilities.

As a result of these perceptions, a steady market exists for poultry products that are grown in a more natural way, such as on pasture. Farmers who take advantage of this public perception can often sell their pasture-based poultry products for premium prices by using well-established language such as "Free-Range" and "Organic".

Q: **What are the real benefits of pasture poultry?**

There are 5 major benefits of raising poultry on pasture:

1. Recent research in Pennsylvania (see American Pasture Poultry Association APPA GRIT! #11) revealed substantial increases in nutritional value of pasture poultry, particularly in Omega-3 Fatty Acids and Vitamin A, plus a significant decrease in total fat.

2. Many people feel that poultry raised on pasture, in fresh air and sunshine, tastes superior to confinement raised poultry. Many also think the naturally raised poultry has a firmer texture and a more satisfying "bite".

3. Small-scale and limited resource farmers can start a profitable farm enterprise for a fraction of the cost of conventional, integrator-controlled poultry housing. Pasture poultry investments as low as only $1,000 can potentially earn a net return of several thousand dollars per year.

4. Moving poultry across the pasture is a way to spread manure and fertility without using excessive equipment, labor or off-farm inputs such as fertilizer.

5. Poultry can be used to scavenge crop residue, and hog down weeds and grasses in multi-crop fields being used for horticulture and floriculture.

Q: **Are there feed savings with pasture poultry?**

No - Experience of many pasture-poultry producers is that 3.5 to 4 pounds of feed are required for each 1 pound of gain. Conventional poultry requires about 2 pounds of feed to get 1 pound of gain. It is entirely possible that pasture poultry requires up to twice the amount of feed as confined poultry.

Q: **How much can I earn with pasture poultry?**

Significantly more than is received by confinement poultry growers. Pasture poultry generally commands higher prices, therefore, the potential for profit is higher than with conventional, confinement poultry. Remember though, that attendant costs are higher, sometimes much higher, than in conventional poultry.

Pasture poultry is usually sold locally, with only minimal processing. Broilers sell for between $1.50 and $3.50 per pound, dressed weight. Turkeys sell for between $2.75 and $3.50 per pound. Prices vary between producers, and between regions of the country. The price difference often depends on whether the birds are sold from the chill tank, or are bagged, weighed, labeled, and ready for the freezer. Production costs are usually about 1/2 to 2/3 the sale price. Many producers do not add their labor into production costs. Pasture raised eggs sell for $1.50 to $3 per dozen.

Q: What does it cost to raise poultry on pasture?

Costs vary between regions and between producers. Pens can range from $100 to $400 depending on size and type. Electrified poultry netting can cost between $1 and $1.50 per lineal foot. Feed costs vary widely across the country, from a low of 8-cents per pound in the Midwest, to as much as 13-cents per pound in the east and west coasts. Organic feed generally costs from 50% to 100% more than conventional feed. Many growers add supplemental vitamins, enzymes, probiotics, and minerals to their feed ration to overcome stress and to help poultry grow better.

Q: What does it cost to process poultry?

Most pasture poultry producers choose to do their own processing. Reasons given include being close to home, earning extra money, and concerns about quality control. The biggest reason many growers do processing, however, is that they don't have a processing facility nearby where they can take their birds. In recent times, almost all local poultry processing houses have gone out of business.

Large integrator processors that only do company birds have replaced them. In some areas, pasture poultry growers have banded together to build Mobile Processing Units. In other cases, growers are choosing to support centralized facilities that are privately owned. Processing equipment costs range from less than $1,000 to more than $20,000, depending on the level of sophistication.

Small-scale commercial poultry facilities that will take small quantities of poultry for processing charge from $1 to $3 per broiler, and from $3 to $8 per turkey.

Q: What methods are used in pasture poultry?

There are FOUR basic methods: PASTURE PENS, CHICKEN TRACTORS, FREE RANGE, and DAY RANGE. Each method has regional refinements:

1. PASTURE PENS are bottomless pens that hold layers, broilers, or turkeys, and are moved daily or as needed to give the poultry fresh pasture. This method was pioneered by Joel Salatin and popularized in his book *PASTURE POULTRY PROFITS*. A typical pen is 10x12 x2-feet, and holds 80 broilers. About 3/4 of the top is roofed; the rest of the top and sides are covered with poultry wire. Each broiler requires 3-inches of feed trough space, and a 10-foot long 6-inch sewer pipe sawed in half lengthwise is a typical feed trough. The birds need a continual supply of fresh water, up to 15 gallons per day per 80 broilers. Some growers use these pens for layers and raise 30 or 40 hens, with nest boxes fixed to the pen side. They reach in from outside the pen to gather eggs. These pens are not well suited for turkey production, although some growers do grow up to 20 turkeys per pen.

2. CHICKEN TRACTOR is a group of methods including free range and Day Range. It is a permaculture design used to acknowledge the whole system that includes the poultry, the garden, and the grower. This idea was first presented by Bill Mollison, and was further popularized by the authors

Andy Lee and Patricia Foreman in their book *CHICKEN TRACTOR: The Permaculture Guide to Happy Hens and Healthy Soil.*

3. FREE RANGE has been practiced for a century or more. This method fell out of favor in the mid-20th century due to disease and predator inroads, and was mostly replaced by commercial confinement poultry production. Free Range generally means a fenced pasture surrounding the barn or poultry shelter. In recent years the term "free range" has lost much of its luster due to USDA definition that simply states "the poultry has access to the outdoors". In this definition no thought is given to how much access, how easy is the access, or whether the "range" is a dust lot or grassy pasture.

4. DAY RANGE is a recent hybridization of chicken tractors and free range. This method was pioneered by Andy Lee and several leading pasture poultry producers across the country. In the Day Range system, the poultry are sheltered at night in mini-barns or portable units that have floors with deep bedding. The floor and bedding keep the birds warm and dry during wet and cold weather. The birds are protected from predators and weather, and allowed to graze in the daytime inside temporary paddocks that are fenced with portable, electric poultry netting. The netting keeps the poultry in, and the predators out.

Q: What are the advantages of Day Range?

1. After they leave the brooder, poultry are housed in a mini-barn, which protects them from weather extremes and predators. This gives them an intermediate hardening off area before going onto cold or wet ground.

2. Because the poultry only use the shelter at night, each shelter can safely handle twice as many poultry as the pasture pens and chicken tractors.

3. Expensive perimeter fencing is not required because the grower keeps the flock in, and predators out, with portable electrified poultry netting.

4. The area for poultry to graze is moved regularly by repositioning the poultry netting. This eliminates over-grazing, and gives the poultry continual access to fresh, growing pasture.

5. In inclement weather, the poultry can be held longer in the shelter thus protecting them from stress and illness.

6. Chores are much faster, since feeding and watering are done in the open fenced area and the shelter is only moved infrequently. Sometimes the shelter is moved weekly, other times the shelter will stay in one location for a month or longer, even for an entire season.

Q: What are the disadvantages of Day Range?

1. This method requires more manager attention to pasture rotations, rather than just methodically moving the pens each day.

2. The shelter is more expensive than the pasture pen, because it has a floor. The floor is covered with bedding to soak up manure. The bedding is cleaned out at the end of each flock, and used for compost. Planer shavings or sawdust are best, since they soak up the moisture from the manure. Keeping the bedding deep enough and dry is important, otherwise it will cake, leading to dirty birds and unhealthy conditions.

3. About 1/3 to 1/2 of the poultry manure is captured in the bedding. This is good from a composting standpoint, but the bedding does require handling, whereas the pasture pen is moved daily, and the manure is spread evenly along the path of travel.

4. Poultry are protected from most predators by the electric poultry netting, but it does not deter flying predators such as hawks. Securing the poultry inside the shelter at night eliminates losses to owls which are nocturnal feeders.

Q: What are the advantages of the pasture pens and chicken tractors?

1. The small shelters are easy and inexpensive to build. When placed on level or gently sloping land, they are relatively easy to move with the aid of a dolly or scoot.

2. The small-scale and easy-to-learn method makes it possible for beginning poultry growers to start and to be successful with limited financial means.

3. The controlled moves will harvest grass and spread manure uniformly across the field.

4. Perimeter fencing is not required, since each flock is contained within the pasture pen.

5. The pasture pens are nearly predator proof. Daily moves keep predators off-balance. Wire sides and heavy wood frames deter predators. Some deaths do result, however, from predators scaring the chickens, causing them to pile up and suffocate. Determined dogs will sometimes get inside these pens, but poultry are safe from flying predators.

Q: What are the disadvantages of pasture pens and chicken tractors?

1. The small pens hold relatively few poultry, compared to their cost.

2. Poultry are removed from the brooder at two to three weeks of age and placed outdoors in minimal shelter. This sometimes results in stress, hypothermia, and frequent mortalities.

3. The two-foot high roof on the pasture pens can trap heat, leading to heat stress that causes losses in weight, and sometimes mortalities. The roof is too low for turkeys to stretch and raise their heads to full height.

4. In rough or hilly land, the heavy pens are difficult to move without injuring the person, or crushing chicks or poults. Recent innovations with PVC pens make the pens lighter and easier to move. These PVC pens are much more ex-

pensive than wooden frame pens, and can blow over in high winds.

5. Pasture pens offer only minimal protection from weather. Even under the best of circumstances the poultry grown in these pens require up to one week longer to mature, and require up to two times as much feed, as confinement raised poultry.

6. Daily chores are time-consuming and can be physically challenging, given the small number of birds serviced at each location.

7. Even though the pasture pens are moved daily, the poultry only has a brief period of fresh graze before the new site is contaminated with manure. And, unless the pen is moved again at dusk, the birds have to bed down in manure-soaked grass. This is unhealthy and unsightly, and leads to dirty feathers, feather loss, and skin sores.

8. The individual pens are hard to move around in market gardens so the poultry can harvest grass and weeds and insects.

Q: What are the advantages of free range?

1. More birds can be held in a given area, and can be taken care of in less time, compared to pasture pens.

2. Poultry are free to move around, and thus able to forage more naturally for grasses, legumes, and bugs. This is especially useful in market gardens where poultry are used to harvest weeds, grasses, and insects, and to spread manure as fertility for the following crop.

3. Poultry are often cleaner, since they don't bed down in manure as in the pasture pens.

Q: What are the disadvantages of free range?

1. Predator losses can decimate entire flocks in a short time.

2. Perimeter fences required to keep in poultry and keep

out predators can be very expensive to install and hard to maintain.

3. Flock pressure is greater on some portions of the site, leading to over-grazing and soil/sod damage. Areas around doorways and near the shelter are often muddy, fecal laden sites, which are ugly and unhealthy.

Q: Why did the chicken cross the pasture?

The proverbial question: "Why did the chicken cross the road" has many answers. Given the subject of this book, we have changed "road" to "pasture". We left out most of the truly tasteless and tacky responses. Credit for these below has to be given to the great minds of many anonymous people who have wondered: "What's the real reason (versus good reason) behind the intent of the chicken crossing the road?"

Oprah — *"They are on their way to Oprah's house for some gourmet low-fat, oven-baked "fried" chicken, Oprah's favorite."*

Martha Stewart — *"This is an Aurcana; its eggs are delicate shaped blue and green whose colors you can find in my custom paint collection at K-Mart."*

Lassie — *"I was hungry and I gave way to chase."*

President Bush — *"We are going to drill for oil in those pastures. The chickens will not be endangered nor disturbed."*

Freud — *"The fact that you thought that the chicken crossed the pasture reveals your underlying insecurity and sexual fantasy about chickens."*

Ernest Hemingway — *"To die. In the rain."*

Saddam Hussein — *"This was an unprovoked act of rebellion and we were quite justified in dropping 50 tons of nerve gas on it."*

Dr. Seuss — *"Did the chicken cross the road? Did he cross it with a toad? Yes the chicken crossed the road, But why he crossed, I've not been told!"*

Jerry Seinfeld — *"Why does anyone cross a pasture? I mean, why doesn't anyone ever think to ask, "What the heck is this chicken doing walking around all over the place anyway?"*

Warren Buffet — *"The chicken knew the pasture has high intrinsic value and was a good deal for long-term investments."*

Charles Darwin — *"Chickens, over great periods of time, have been naturally selected in such a way that they are now genetically predisposed to cross pastures."*

Richard M. Nixon — *"The chicken did not cross the pasture. I repeat, the chicken did not cross the pasture. Can anyone find that chicken on tape?"*

Oliver Stone — *"The question is not "Why did the chicken cross the pasture?" But is rather "Who was crossing the pasture at the same time whom we overlooked in our haste to observe the chicken crossing?"*

Martin Luther King, Jr. — *"I envision a world where all chickens will be free to cross pastures without having their motives called into question."*

Bill Gates — *"I have just released the new Chicken 2002, which will both cross pastures AND balance your checkbook, although when it divides 3 by 2 it gets 1.4999999999."*

Jabez — *"The lord blessed me indeed and enlarged my pasture. Thy hand was with me, and kept me from evil factory farms."*

M.C. Escher — *"That depends on which plane of reality the chicken was on at the time."*

George Orwell — *"Because the government had fooled them into thinking that they were crossing the pasture of their own free will, when they were really only serving their interests."*

Colonel Sanders — *"Damn! I missed one?"*

Plato — *"For the greater good."*

Homer Simpson — *"Mmmmmrn. . . chicken."*

Aristotle — *"To actualize its potential."*

Dalai Lama — *"Pastures never wait but keep growing, correspondingly, our lives keep growing onward all the time. We cannot turn back time. In that sense, there is no genuine second chance to recross a pasture. Don't waste precious moments of life no matter what pasture you are in."*

Karl Marx — *"It was a historical inevitability. Pastures are the opium for all chickens."*

Nietzsche — *"Because if you gaze too long across the pasture, the pasture gazes also across you."*

B.F. Skinner — *"Because the external influences, which had pervaded its sensorium from birth, had caused it to develop in such a fashion that it would tend to cross pastures, even while believing these actions to be of its own free will."*

Carl Jung — *"The confluence of events in the cultural gestalt necessitated that individual chickens cross pastures. This brought such occurrences into being in their quest for the creation of each chicken's personal mandala."*

John Locke — *"Because he was exercising his natural right to liberty."*

Albert Camus — *"It doesn't matter; the chicken's actions have no meaning except to him."*

Jean-Paul Sartre — *"In order to act in good faith and be true to itself, the chicken found it necessary to cross the pasture."*

Albert Einstein — *"Whether the chicken crossed the pasture or the pasture crossed the chicken depends upon your frame of reference."*

Grandpa — *"In my day, we didn't ask why the chicken crossed the pasture. Someone told us that the chicken had crossed the pasture, and that was good enough for us."*

Carolyn Myss — *"A new energetic species of chicken is evolving, but it has to cross many pastures to gain awareness of when it's there."*

Ralph Waldo Emerson — *"It didn't cross the pasture; it transcended it."*

O.J. — *"It didn't. I was playing golf with it at the time. Why, I don't even own a chicken!"*

James Redfield — *"To cross the pasture and stop the polizarization of Fear each of us must participate personally. We must watch our thoughts and catch ourselves every time we treat another as an enemy."*

Machiavelli — *"The point is that the chicken crossed the pasture. Who cares why? The ends of crossing the pasture justify whatever motive there was."*

Victor E. Frankl — *"Most pastures are in bad shape, but everything will become still worse unless each of us does our best. So, let us be alert - alert in a two fold sense: Since Auschwitz we know what man is capable of. And since Hiroshima we know what is at stake."*

Timothy Leary — *"Because that's the only kind of trip the Establishment would let it take."*

Fox Muider (the guy from the X-flies) — *"It was a government conspiracy."*

Immanuel Kant — *"The chicken, being an autonomous being, chose to cross the pasture of his own free will."*

Joseph Stalin — *"I don't care. Catch it. I need its eggs to make my omelet."*

Thomas de Torquezuada — *"Give me ten minutes with the chicken and I'll find out."*

Terrence McKenna — *"History Ends in Green — there's mushrooms in them there green pastures."*

Joel Salatin — *"I had to pull and pull to get that chicken to cross that pasture, now it's a fertile salad-bar sea of green."*

Andy Lee — *"The chicken crossed the road in a chicken tractor. Never walk when you can ride, even across the pasture."*

Patricia Foreman — *"The chicken crossed the pasture as a metaphor to "retire early and often"–there are many pastures I have yet to cross. Why? Just for the knee-slapping fun of it. I've only just begun to be put-out-to-pasture."*

Resource Guide

Poultry Producers & Equipment Dealers

Cackle Hatchery, PO Box 529, Lebanon, MO 65536, 417-532-4581, www.cacklehatchery.com

Clausing Company, Nocatee, FL 34268, 941-993-2542

Clearview Hatchery, PO Box 399, Gratz, PA 17030, 717-365-3234

Cutler's Supply, Inc., 1940 Old 51, Applegate, MI, 48401, 810-633-9450, Email: sales@cutlersupply.com, www.cutlersupply.com

Fairview Hatchery, 18795 S. 580 W., Remington, IN, 47977, 800-440-1530, Fax (219)261-2197, Email: sales@fairviewhatchery.com, www.fairviewhatchery.com

Hoffman Hatchery, PO Box 129, Gratz, PA 17030, 717-365-3694

Inman Hatcheries, PO Box 616, Aberdeen, SD 57402, 605-225-8122, www.hoffmanhatchery.com

Marti's Hatchery, PO Box 27, Windsor, MO 65360, 816-647-3156

Murray McMurray Hatchery, PO Box 458, Webster City, IA, 50595, 515-832-3280. These folks have a beautiful full color catalogue.

Sand Hill Preservation Center, 1878 230th St, Calamus, IA, 52729, 563-246-2299, www.sandhillpreservation.com

Southwest Poultry Supply, Wayne Holly, 500 Honey Creek Dr., Southwest City, Missouri, 64863, 417-762-3201, Email: SWPoultrySupply@hotmail.com

Stork Hatchery and Farm Store, Box 213, Fredericksburg, IA, 50630, 319-237-5981

Wish Poultry, 802 South Hall, Box 862, Prairie City, OR, 97869, 541-820-3509

Shank's Hatchery, PO Box 429, Hubbard, OR, 97032, 503-981-7801

Carolyn Christman at American Livestock Breeds Conservancy, tells us many of the above hatcheries have one or more rare varieties of turkeys available.

For more information about heritage and minor breed poultry, in USA contact American Livestock Breeds Conservancy, Phone 919-542-5704, Website: www.albc-usa.org

For information on hatcheries in Canada, contact Rare Breeds Canada, 341-1 Clarkson Road, RR# 1 Castleton ON, K0K 1M0, Phone: 905-344-7768, Email: rbc@rarebreedscanada.ca Website: www.rarebreedscanada.ca

Nutritional Supplements & Soil Amendments

Countryside Natural Products — They stock most natural feed supplements and soil treatments. They also have an organically certified dealership for grains. PO Box 997, Fishersville, VA 22939, Order 888-699-7088 or 540-932-8534

Fertrell Company — This is a major supplier of nutritional supplements and soil amendments. PO Box 265, Bainbridge, Pennsylvania 17502, 717-367-1566

North Country Organics — All natural land care products PO box 372, Bradford, Vermont 05033, 802-222-4277 www.norganics.com

Redmond Minerals, Inc. — These folks mine the natural rock salt. 6005 North 100 West, PO Box 219 Redmond, Utah 84652, Order Natural Rock Salt 800-367-7258

Seven Springs Farm — Organic meats, farming, & gardening supplies, 426 Jerry Lane NE - Check, VA 24072, Located in Floyd County Virginia, 540-651-3228 or 800-540-2981, E-mail: 7springs@swva.net, www.7springsfarm.com

Thorvin Kelp Call for nearest dealer 800-464-0417

Associations to Join

American Livestock Breeds Conservancy, is one of our favorite non-profit organizations. They are dedicated to preserving America's heritage livestock. Don Bixby, DVM, is the current director. You can reach them at PO Box 477, Pittsboro NC 27312, 919-542-5704. www.albc-usa.org

The All-American Turkey Growers Association (AATGA) is an old organization reborn. Originally founded in the mid 1920's in Grand Forks, North Dakota, the AATGA established many of the standards for turkeys. The AATGA waned with the decline of the small-scale breeding and sale of turkeys. The AATGA promises to "offer its every help to those seeking standard turkeys once again". They place a high priority on serving youth organizations like Future Farmers of America and 4-H. They are considering the revival of live-breeder bird classes and live market classes. They publish a quarterly bulletin. Norm Kardosh is the main force behind this group. For membership information contact: AATGA, Good Sheperd Ranch, 730 Smoky Valley Road, Lindsborg, KS 67456, Phone (785) 227-3972.

The Standard Turkey Preservation Association (STPA) was started in 1999 by Sheane and Bonnie Meikle. The STPA was formed to link breeders for rare and heritage turkeys and to educate the public about turkey related issues. To find out more or become a member contact STPA, Box 7 Site 6, RR#2, Ponoka, AB T4J IR2, CANADA, (403) 783-6632, standard_turkey@hotmail.com.

World Wide Web & Internet Resources

Information on a wide range of livestock and poultry topics is available from:

www.attra.org/search.html

Chicken health:

www.msstate.edu/dept/poultry

Chicken breed listings offering general information:

www.urbanext.uiuc.edu/eggs/res10-breedhistory.html#2 oultry

Day Range Page 296

Periodicals and Internet Discussion Groups

pasturepoultry@yahoogroups.com, an authoritative discussion by those who are doing it. Archives available with valuable information and insights. Moderated by Andy Lee.

APPPA GRIT!, the quarterly newsletter of the 500-member American Pasture Poultry Producers Association, c/o Diane Kaufmann, 5207 70th St, Chippewa Falls, WI 54729. $20 annual membership dues, includes newsletter.

SMALL FARM TODAY, Magazine regularly carries articles on pasture poultry and small-scale sustainable agriculture. $21 year, 6 issues. 3903 West Ridge Trail Road, Clark, MO 65243, phone 800-633-2535. Has an excellent book catalog. www.smallfarmtoday.com

ACRES USA, 12 issues, $24. PO Box 91299, Austin, TX 78709, phone 512-892-4400. Covers wide range of sustainable agriculture topics. Has an excellent book catalog.

STOCKMAN GRASS FARMER, 12 issues $28. PO Box 2300, Ridgeland, MS 39158. 800-748-9808. The grazer's voice, regular in-depth articles on pasture based poultry and livestock systems. Has an excellent book catalog. www.stockmangrassfarmer.com

WATT Poultry USA, 122 S. Wesley Ave, Mt. Morris Illinois 61054-1497. 815-734-4171. The voice of the commercial poultry industry, contains articles relative to growing, processing, and marketing poultry products nationally and internationally. www.wattnet.com

Processing Equipment Suppliers

M & M Poultry Equipment, Lawrenceville, GA, or South Hwy. 65, Hollister, MO, 417-334-6641

Ashley Equipment, PO Box 2, Greensburg, IN 47240, 812-663-2180

Brower Equipment, PO Box 2000, Houghton, IA 52631, 319-469-4141 or 800-553-1791, www.browerequip.com

R and R Pluckers, Rob Bauman, RD 1, Oxford, NY 13830, 607-843-7415

Pickwick Zesco, 7887 Fuller Road Suite 116, Eden Prairie, MN 55344, 800-808-3335

JAKO Inc, 6003 E. Eales Rd., Hutchinson, KS 67501, 877-525-6462, www.jakoinc.com

Zesco, 7887 Fuller Road, Suite 117, Eden Prairie, MN 55344, Phone 800-808-3335.

Other Supplies & Equipment

G&M Sales (nest liners, tanks), 5462 S. Valley Pike, (Route 11), Harrisonburg, VA 22801, 800-296-9156, Email: rreeves@gmsalesofva.com, www.GMSalesofVA.com

Koch Supplies (bags, supplies, food processing), Kansas City, MO, 800-777-5624

Matthiesen Company (bags, supplies), San Antonio, TX 800-624-8635, www.matthiesenequipment.com

Cryovac Sealed Air Corporation (waterproof, shrink wrap bags), Duncan, SC, 800-845-3456, www.cryovac.com

Kencove Farm Fence, Inc. (fencing supplies), Catalog ordering 1-800-536-2683, 24 hour Fax Ordering 724-459-9148, www.kencove.com

Premier1Supplies.com (fencing supplies), 800-282-6631, Email: info@premier1supplies.com, www.Premier1Supplies.com

Hatchery Equipment.com, HawkHead International, Inc, 200 Industrial Loop #158, Orange Park, FL 32073, 904-264-4295, Fax:904-264-6899, 800-252-4295, Email:sales@hawkhead.com, www.HatcheryEquipment.com

Farm Tek's Tek Supply Catalog (the "Chicken Inn"–a Day Range Shelter in a kit), 800-327-6835, www.FarmTek.com

Featherman Portable Shelters, David Schafer, 760 SW 55th Ave, Jamesport, MO 64648, 660-684-6035, Email: sales@schaferfarmsnaturalmeats.com, www.schaferfarmsnaturalmeats.com

NASCO, Retail locations: Fort Atkinson, WI and Modesto, CA, Order by phone at 800-558-9595, View online catalog at www.eNasco.com

Bibliography

Many of the books we used in researching *Day Range Poultry* are from foreign presses, small presses or are out-of-print. We found many of the books below through searches on the world wide web. You can use your search engines to find used bookstores. Also check the following sites.

www.bookfinder.com
www.alibris.com
www.bibliofind.com

AGRICulture OnLine Access (AGRICOLA) – the Department of Agriculture's Library search at: INK http://www.nal.usda.gov/ag98/english/catalog-basic.html

Baker, Helen, *Turkeys: Common Sense Theories, Practical Management, Incubation and Brooding in Detail, Feeding Directions, Feeding Formulas*. 1933. Self-published; out-of-print.

Christman, Carolyn and Robert Hawes, *Birds of a Feather: Saving Rare Turkeys from Extinction*, American Livestock Breeds Conservancy, Pittsboro, North Carolina, USA, 1999.

Cline, L. E., *Turkey Production: Complete Text on Breeding, Feeding, Handling, Marketing and Disease Control*, Orange Judd, New York, 1929, out-of-print.

Damerow, Gail, *Chicken Health Handbook*, Storey Communications, Pownal, Vermont, USA, 1994.

Etches, Robert, *Reproduction in Poultry*, University of Guelph, Ontario, Canada, 1996.

Feltwell, Ray, *Small-Scale Poultry Keeping: A Guide to Free-Range Poultry Production*, Faber and Faber, Boston, USA, 1980 & 1992.

Feltwell, Ray, *Turkey Farming*, Faber and Faber, London, 1953, out-of-print.

Hartman, Roland C., *Hatchery Management*, Orange Judd, New York, 1951, out-of-print.

Harvey, Rob, *Practical Incubation*, Hancock House Publishers, Washington, USA, 1990.

Lamon, Harry and Rob Slocum, *Turkey Raising*, Orange Judd Publishing, 1927, out-of-print.

Marsden, Stanley and J. Holmes Martin, *Turkey Management*, Interstate, Danville, IL, First Edition 1939 and 5th edition, 1955, out of-print.

Lee, Andy, *Backyard Market Gardening: The Entrepreneur's Guide to Selling What You Grow*, Good Earth Publications, Lexington, Virginia 24416, 1993.

Lee, Andy and Patricia Foreman, *Chicken Tractor: The Permaculture Guide to Happy Hens and Healthy Soil,* Good Earth Publications, Lexington, Virginia 24416, Second Edition 1994.

Lewis, Harry R.M., *Productive Poultry Husbandry*, Lippincott's Farm Manuals, Lippincott Company, Chicago, IL, 1933 – Out-of-print.

Mollison, Bill, *Permaculture: A Designer's Manual*, Tagari Publications, Tyalgum, Australia, 576 pages, 1988.

Price, Weston, *Nutrition and Physical Degeneration*, Keats Publishing, Los Angeles, 1997.

Roberts, Michael, *Turkeys at Home*, Domestic Fowl Trust, England, 1989.

Roberts, Michael, *Modern Free Range*, Domestic Fowl Research, England, 1988.

Rose, S.P., *Principles of Poultry Science*, CAB International, UK, 1997.

Salatin, Joel, *Pasture Poultry Profit$: Net $25,000 in 6 Months on 20 Acres*, Polyface, Inc, Swoope, Virginia, 1993.

Salatin, Joel, *Salad Bar Beef*, Polyface, Inc, Swoope, Virginia, 1995.

Salatin, Joel, *You Can Farm: The Entrepreneur's Guide to Start and Succeed in a Farming Enterprise*, Polyface, Inc, Swoope, Virginia, 1998.

Stromg, Loyl, *Sexing all Fowl, Baby Chicks, Game Birds, and Cage Birds*, Stromberg Publishing, Pine River, Minnesota, USA, 1997.

Stromberg, Janet, *A Guide to Better Hatching*, Stromberg Publishing, Pine River, Minnesota, 1975.

Thear, Katie, *Free-range Poultry*, Farming Press Books, Ispwich, UK, 1990.

Thear, Katie, *Incubation: A Guide to Hatching and Rearing*, Broad Leys Publishing Co, London, 1997.

Yeomans, P.A., *Water for Every Farm: Using the Keyline Plan*, Second Row Press, Katoomba, Australia, 1981.

Index

The End...
And the Beginning...

Good Earth Publications, Inc.
Established 1989

For Additional Titles
Visit Our Web Site and Book Store at:
www.GoodEarthPublications.com

A Tiny Home to Call Your Own
Living Well in Just Right Houses
You can live in an attractive, aesthetically appealing, upscale house and you can
have a home that is quality built, architecturally beautiful,
highly marketable, and profitable.
by Patricia Foreman & Andy Lee, 208 pages, $19.95

Chicken Tractor
The Permaculture Guide to Happy Hens & Healthy Soil
This is a revolutionary, practical, hands-on book for gardeners and poultry growers. It has already helped thousands of gardeners have better gardens and has
literally changed the lives of millions of chickens all over the world.
by Andy Lee & Patricia Foreman, 320 pages, $22.95

Backyard Market Gardening
The Entrepreneur's Guide to Selling What You Grow
Discover how easy and profitable it is to grow and sell vegetables, fruits, flow-
ers, herbs, and small livestock from your own Backyard Market Garden.
by Andy Lee & Patricia Foreman, 352 pages, $19.95

ORDER AT:
www.GoodEarthPublications.com
Phone (540)261-8874 or 800-499-3201

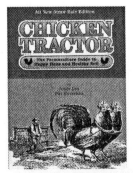

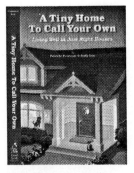

Printed in the United States
41591LVS00004B/146

9 780962 464874